Study Guide and Practice Workbook

Mathematics
Applications and Connections

Course 1

Glencoe
McGraw-Hill

New York, New York   Columbus, Ohio

To the Student

This *Study Guide and Practice Workbook* gives you additional examples and practice for the concept exercises in each lesson. The exercises are designed to aid your study of mathematics by reinforcing important mathematical skills needed to succeed in the everyday world. The material is organized by chapter and lesson, with one study guide worksheet and one practice worksheet for every lesson in *Mathematics: Application and Connections*, Course 1.

Always keep your workbook handy. Along with your textbook, daily homework, and class notes, the completed *Study Guide and Practice Workbook* can help you in reviewing for quizzes and tests.

To the Teacher

Answers to each worksheet are found in the *Study Guide Masters* or *Practice Masters* booklets and also in the Teacher's Wraparound Edition of *Mathematics: Applications and Connections*, Course 1.

Glencoe/McGraw-Hill

A Division of The McGraw·Hill Companies

Send all inquiries to:
Glencoe/McGraw-Hill
8787 Orion Place
Columbus, OH 43240

ISBN: 0-02-833123-0

Study Guide and Practice Workbook, Course 1

22 23 24 25   DOH   15 14 13

Contents

1-1

Study Guide

A Plan for Problem Solving

A one-year subscription to a popular magazine costs $24. Ted is thinking of sharing the cost of the subscription equally with two friends. How much will each have to pay?

Explore	What do you know? A one-year subscription costs $24. Ted will share the cost equally with two friends. What are you trying to find? How much each friend will have to pay.
Plan	Since Ted will share with two friends, there are 1 + 2 or 3 people sharing the cost. Find the number of times 3 goes into $24.
Solve	$24 ÷ 3 = $8 Each friend will pay $8 toward the subscription.
Examine	The answer makes sense. 3 × $8 = $24

Use the four-step plan to solve each problem.

1. A record weight for a red snapper was 46 pounds. A record weight for a sturgeon was 468 pounds. About how many times heavier than the red snapper was the sturgeon?

2. The distance from Chicago to Cleveland is 344 miles. Lisa plans to drive this distance in two days. If she drives 180 miles today, how many miles will she have to drive tomorrow?

3. On a certain day, 525 people signed up to play softball. If 15 players are assigned to each team, how many teams can be formed?

4. The Farrells want to buy a VCR that costs $460. They plan to make a down payment of $100 and pay the rest in eight equal payments. What will be the amount of each payment?

5. Josita received $50 as a gift. She plans to buy two cassette tapes that cost $9 each and a deluxe headphone set that costs $25. How much money will she have left?

6. Tran wants to buy a bicycle that costs $192. He plans to save $6 each week. How many weeks will it take him to save enough money?

1-1 Practice

A Plan for Problem Solving

Use the four-step plan to solve each problem.

1. Kara is a member of the school swim team. She swims 35 laps at each practice. The swim team practices 4 days per week. How many laps does Kara swim in four weeks?

2. The school band is having a car wash to raise money to purchase new uniforms. If they charge $4 per car, how many cars must band members wash to raise $208?

3. When the Breen family left for their vacation, the car's odometer read 5,289 miles. Upon their return, the odometer read 6,035 miles. How many miles did the Breen family travel during their vacation?

4. Kyle is renewing his subscription to his favorite computer magazine. The cost is $24 for 12 issues. What is the cost of each issue?

5. The sixth-graders at Danville Middle School collected aluminum cans for a recycling project. Room 6A collected 137 cans; Room 6B collected 98 cans; Room 6C collected 164 cans. How many aluminum cans were collected in all?

6. Carla enjoys collecting baseball cards. Her album contains 115 pages. Each page holds 9 baseball cards. How many baseball cards will there be in her album when it is full?

7. Chi-Sun went shopping for hiking supplies. He bought a knapsack for $22, a canteen for $5 and a compass for $12. How much money did he spend in all?

8. Eva has saved $64 from her babysitting money. She needs 4 times this amount to buy a new bike. How much does the new bike cost?

Name _____ **Date** _____

1-2 Study Guide

Using Patterns

At the right is a pattern of geometric figures. What is the area of the eighth figure in the pattern? Counting the square units, you see a number pattern: add 2 to each number. So, to find the area of the eighth figure, continue the number pattern until you come to the eighth number.

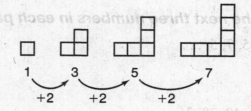

1 — 3 — 5 — 7 — 9 — 11 — 13 — 15
  +2   +2   +2   +2   +2   +2   +2

The area of the eighth figure is 15 square units.

Example   **Find the next three numbers for the pattern.**
1, 3, 9, 27, _____, _____, _____

Study the pattern. How do you get each succeeding number?

1, 3, 9, 27, _____, _____, _____
  ×3 ×3 ×3

Each number is 3 times the number before it.
1 × 3 = 3   3 × 3 = 9   9 × 3 = 27

The next three numbers are 27 × 3 or 81, 81 × 3 or 243, and 243 × 3 or 729.

Find the next three numbers in each pattern.

1. 27, 23, 19, 15, _____, _____, _____

2. 27, 35, 43, 51, _____, _____, _____

3. 64, 32, 16, 8, _____, _____, _____

4. 2, 10, 50, 250, _____, _____, _____

Draw the next three figures in each pattern.

5.

6.

Mathematics: Applications and Connections, Course 1

1-2 Practice

Using Patterns

Find the next three numbers in each pattern.

1. 7, 5, 7, 5, . . .

2. 2, 5, 8, 11, . . .

3. 42, 40, 38, 36, . . .

4. 1, 4, 16, 64, . . .

5. 1, 9, 6, 1, 9, . . .

6. 5, 10, 15, 20, . . .

7. 192, 96, 48, 24, . . .

8. 1, 2, 4, 7, . . .

9. 1, 10, 2, 20, . . .

10. 1, 4, 9, 16, . . .

11. 17, 16, 13, 12, . . .

12. 8, 27, 64, . . .

Draw the next two figures in each pattern.

13.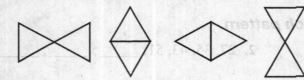

14.

15.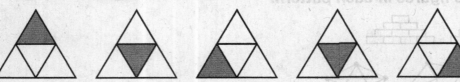

Mathematics: Applications and Connections, Course 1

1-3 Study Guide

Estimation by Rounding

You can use these rules to round whole numbers.

Rules for Rounding
Look at the digit to the right of the place being rounded. • The digit remains the same if the digit to the right is 0, 1, 2, 3, or 4. • Round up if the digit to the right is 5, 6, 7, 8, or 9.

Examples

1 Round 6,905 to the nearest ten, to the nearest hundred, and to the nearest thousand.

6,9_0_5	5 = 5, round 0 up to 1.	6,910
6,_9_05	0 < 5, leave 9 unchanged.	6,900
_6_,905	9 > 5, round 6 up to 7.	7,000

2 Round $48.29 to the nearest dollar and to the nearest ten dollars.

| $4_8_.29 | 2 < 5, leave 8 unchanged. | $48.00 |
| $_4_8.29 | 8 > 5, round 4 up to 5. | $50.00 |

Round each number to the underlined place-value position.

1. _3_92

2. _4_7

3. _1_,013

4. _6_74

5. _7_31

6. 8,2_0_4

7. _3_,789

8. _6_06

9. 4,1_1_1

10. _7_,619

11. 2,_5_88

12. 5,0_4_5

Estimate. State whether the answer shown is reasonable.

13. $8.90 + $0.45 + $1.25 = $13.72

14. 812 − 193 = 619

15. $7.89 × 3 = $23.67

16. 1,164 ÷ 582 = 3

1-3 Practice

Estimation by Rounding

Round each number to the underlined place-value position.

1. 4̲4

2. 6̲5

3. 2̲8

4. 6̲17

5. 5̲93

6. 4̲52

7. 7̲05

8. 6̲99

9. 8̲49

10. 1̲09

11. 68̲4

12. 27̲1

13. 3̲,150

14. 9̲,738

15. 5̲,426

16. 4̲,549

17. 6,6̲55

18. 9,96̲7

19. 1,36̲9

20. 7̲,510

21. 4̲,087

22. 5̲,435

23. 2,90̲3

24. 6,25̲8

25. 3,0̲09

26. 7,49̲9

27. 9̲,501

Estimate. State whether the answer shown is reasonable.

28. 37 + 55 + 81 = 173

29. 28 + 72 + 15 + 129 = 244

30. 562 + 349 + 107 = 1,218

31. 418 + 689 + 104 = 1,411

32. $8.65 + $3.39 + $4.52 = $19.56

33. 976 − 453 = 523

34. 758 − 364 = 394

35. $4.11 + $16.75 + $10.95 = $29.11

1-4 Study Guide

Order of Operations

To evaluate expressions, use the order of operations.

Order of Operations
1. Multiply and divide in order from left to right.
2. Add and subtract in order from left to right.

Example Find the value of $2 \times 8 - 42 \div 7 + 4$.

$$\begin{aligned} 2 \times 8 - 42 \div 7 + 4 &= 16 - 42 \div 7 + 4 \\ &= 16 - 6 + 4 \\ &= 10 + 4 \\ &= 14 \end{aligned}$$

Multiply 2 and 8.
Divide 42 by 7.
Subtract 6 from 16.
Add 10 and 4.

Find the value of each expression.

1. $15 + 4 \times 5 - 6$

2. $40 \div 8 - 3$

3. $9 \times 3 + 3$

4. $20 - 15 \div 3$

5. $2 \times 4 \times 5$

6. $18 \div 2 + 9 \times 2$

7. $3 + 7 - 5 + 2$

8. $4 \times 9 - 9 \times 4$

9. $15 \div 3 + 3 \times 5$

10. $45 \div 9 \div 5$

11. $18 + 3 + 15 \div 3$

12. $50 - 5 \times 5 - 25$

13. $35 \div 7 - 5$

14. $12 \times 2 - 5 \times 4$

15. $4 + 18 \div 3$

16. $150 \div 10 - 3 \times 5$

17. $7 + 2 \times 5$

18. $3 \times 8 - 16$

19. $80 \times 2 \div 40 - 1$

20. $25 - 3 \times 4$

21. $2 + 4 \times 9 \div 12$

22. $45 - 40 + 1 \times 2$

23. $18 \div 2 \times 9$

24. $2 \times 5 \times 3$

1-4 Practice

Order of Operations

Name the operation that should be done first. Then find the value of the expression.

1. $17 - 2 \cdot 4$

2. $24 \div 6 + 3$

3. $11 + 9 - 12$

4. $30 \div 6 \cdot 5$

5. $13 - 4 + 7$

6. $8 \times 7 + 2$

7. $6 \cdot 7 - 2$

8. $36 \div 9 - 3$

9. $24 \times 2 \div 6$

Find the value of each expression.

10. $2 + 4 \times 8$

11. $12 - 9 \div 3$

12. $8 \cdot 7 - 4$

13. $10 + 25 \div 5$

14. $6 + 24 \div 3 \times 2$

15. $15 - 9 \cdot 4 \div 12$

16. $8 \cdot 9 \div 6 - 4$

17. $54 \div 6 - 3 \times 2$

18. $2 \cdot 4 + 16 \div 8$

19. $6 + 21 \div 3 - 9$

20. $12 - 5 \times 6 \div 3$

21. $2 + 7 \cdot 8 \div 4$

22. $63 \div 9 + 12 - 2$

23. $3 \times 6 + 14 \div 2$

24. $7 \cdot 4 - 3 \cdot 8$

25. $9 + 18 \div 3 \times 5$

26. $7 + 8 - 3 + 5$

27. $7 \cdot 5 + 2 \cdot 3$

28. $20 - 5 \cdot 2$

29. $24 \div 8 + 2$

30. $18 + 24 \div 12 + 5$

31. $6 + 5 \cdot 2 + 4$

32. $75 \div 15 \cdot 4$

33. $45 + 22 \div 11$

34. $32 \cdot 4 \div 2$

35. $9 + 4 - 2 \times 3$

36. $39 - 9 \cdot 3 + 6$

37. $15 \times 2 + 5 \times 6$

38. $10 + 53 - 60$

39. $32 - 4 \times 2 + 10$

1-5 Study Guide

Integration : Algebra
Variables and Expressions

The distance around a rectangle is two times its length plus two times its width.

If we use the letter ℓ to represent length and the letter w to represent width, the distance around a rectangle may be written as the algebraic expression $2\ell + 2w$.

ℓ and w are variables. They change for different rectangles.

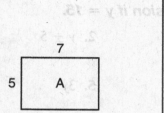

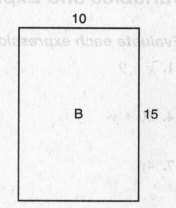

Example   **Find the distance around each of the rectangles shown.**

Rectangle A
$\ell = 5$ $w = 7$

Evaluate $2\ell + 2w$.
$2 \times 5 + 2 \times 7 =$
$\quad 10 \quad + \quad 14 \quad = 24$

Rectangle B
$\ell = 15$ $w = 10$

Evaluate $2\ell + 2w$.
$2 \times 15 + 2 \times 10 =$
$\quad 30 \quad + \quad 20 \quad = 50$

Evaluate each expression if $a = 4$ and $b = 8$.

1. $a + 8$

2. $b + a$

3. ba

4. $b \div a$

5. $3b$

6. $48 \div b$

7. $a - 3$

8. $a \times 0$

9. $4a - b$

10. $5 + b - 3$

11. $b \div a + 8$

12. $4b - 30$

Evaluate each expression if $\ell = 3$, $m = 6$, $n = 12$.

13. $n \div 4 + \ell$

14. $m\ell + n$

15. $n \div m\ell$

16. $n + 3 \times 2$

17. $m\ell - 3$

18. $5m$

19. ℓm

20. $2\ell + m$

Mathematics: Applications and Connections, Course 1

1-5 Practice

Integration: Algebra
Variables and Expressions

Evaluate each expression if y = 15.

1. $y - 9$

2. $y \div 5$

3. $6 + y$

4. $30 \div y$

5. $3y$

6. $28 - y$

7. $4y$

8. $y + 8$

9. $y \div 3$

10. $80 - y$

11. $y - 2$

12. $90 \div y$

Evaluate each expression if r = 6 and s = 8.

13. $3r$

14. $48 \div s$

15. $r + s$

16. $15 - s$

17. $s - r$

18. $s \div 2$

19. sr

20. $r + 5$

21. $4s$

22. $64 \div s$

23. $5r$

24. $s + 16$

Evaluate each expression if j = 2, k = 7, and m = 9.

25. $k + m$

26. $j \times 6$

27. $54 \div m$

28. $k - j$

29. $21 \div k$

30. km

31. $4k \div j$

32. $2m - 5j$

33. $3m + j$

34. $42 \div k + m$

35. $j + m - k$

36. $jk + 7$

37. $2j + k - m$

38. $km + 7$

39. $jm - k$

Mathematics: Applications and Connections, Course 1

1-6 Study Guide

Integration: Algebra
Powers and Exponents

A **power** can be used to show a number multiplied by itself.

Examples

1 Write $a \cdot a \cdot a \cdot a$ as a product using exponents.

$a \cdot a \cdot a \cdot a$ can be written as a^4.

It is read "a to the fourth power."

The exponent, 4, tells how many times the base, a, is used as a factor.

base $\longrightarrow a^4 \longleftarrow$ exponent

2 Write $10 \cdot 10 \cdot 10 \cdot 5 \cdot 5 \cdot 5$ as a product using exponents.

10 is used as a factor 3 times.

5 is used as a factor 3 times.

$10 \cdot 10 \cdot 10 \cdot 5 \cdot 5 \cdot 5 = 10^3 \cdot 5^3$

Multiply to evaluate expressions with exponents.

Examples

3 Evaluate 7^3.

$7^3 = 7 \cdot 7 \cdot 7$
$= 343$

4 Evaluate $2^3 \cdot 8^2$

$2^3 \cdot 8^2 = 2 \cdot 2 \cdot 2 \cdot 8 \cdot 8$
$= 8 \cdot 64$
$= 512$

Write each product using exponents.

1. $6 \cdot 6 \cdot 6 \cdot 6 \cdot 6 \cdot 6$

2. $k \cdot k \cdot k \cdot k$

3. $5 \cdot 5 \cdot 5 \cdot 5 \cdot 5 \cdot 5 \cdot 5$

4. $2 \cdot 2 \cdot 2 \cdot 6 \cdot 6$

5. $c \cdot c \cdot c \cdot c \cdot d \cdot d$

6. $1 \cdot 1 \cdot 1 \cdot 1 \cdot 7 \cdot 7$

Write each power as a product.

7. y^3

8. 15^5

9. $r^3 \cdot s^6$

10. 9^6

11. 100^3

12. 120^2

Evaluate each expression.

13. 2^5

14. 3^3

15. 10^6

16. $2^2 \cdot 4^3$

17. $7^1 \cdot 10^3 \cdot 1^8$

18. 9 squared

1-6 Practice

Integration: Algebra
Powers and Exponents

Write each product using exponents.

1. $5 \cdot 5$

2. $m \cdot m \cdot m \cdot m$

3. $6 \cdot 6 \cdot 6 \cdot 6 \cdot 6 \cdot 6$

4. $a \cdot a \cdot a$

5. $8 \cdot 8 \cdot 8 \cdot 8 \cdot 8$

6. $11 \cdot 11$

7. $4 \cdot 4 \cdot 2 \cdot 2 \cdot 2$

8. $c \cdot c \cdot c \cdot c \cdot d \cdot d$

9. $1 \cdot 1 \cdot 1 \cdot 10 \cdot 10$

Write each power as a product.

10. 3^8

11. 12^6

12. n^7

13. 46^5

14. x^3

15. 78^4

16. 139^3

17. 806^4

18. $y^2 \cdot z^4$

Evaluate each expression.

19. 8^4

20. 0^{10}

21. 4^5

22. 9 cubed

23. $1^8 \cdot 2^4 \cdot 7^2$

24. $12^1 + 4^3 + 3^5$

25. $2^4 + 6 \cdot 3^2$

26. $3^3 \cdot 6^2 \cdot 5^4$

27. 100 squared

28. Evaluate p^6 if $p = 3$.

29. Evaluate s^3 if $s = 8$.

30. Evaluate r^2 if $r = 17$.

Study Guide

Integration: Algebra
Solving Equations

An **equation** is a mathematical sentence that contains an equals sign.
The **solution** is the number that makes the equation true.

Example

Warren spent $120 for theater tickets. If each ticket cost $20, how many tickets did Warren buy?

Let t equal the number of tickets.
The problem may be represented by this equation: $\$20 \times t = \$120$.

Guess and check.

Replace t with 7.	Replace t with 6.
$\$20 \times t = \$120$	$\$20 \times t = \$120$
$\$20 \times 7 = \$120$	$\$20 \times 6 = \$120$
$\$140 = \$120$	$\$120 = \$120$ ✔
The sentence is false.	The sentence is true.

The solution is 6. Warren bought 6 tickets.

Identify the solution to each equation from the list given.

1. $15 + r = 21$ 5, 6, 7

2. $y \div 6 = 7$ 42, 48, 56

3. $5m = 40$ 6, 7, 8

4. $n = 7 + 21$ 3, 14, 28

5. $9 = 63 \div b$ 7, 8, 9

6. $5w = 10$ 2, 25, 50

Solve each equation mentally.

7. $y = 10 + 6$

8. $p = 18 - 9$

9. $x \div 6 = 4$

10. $h \div 5 = 25$

11. $3t = 15$

12. $a - 7 = 16$

13. $18 = 9 + v$

14. $28 = 4p$

1-7 Practice

Integration: Algebra
Solving Equations

Tell whether the equation is true or false by replacing the variable with the given value.

1. $a - 53 = 32; a = 85$

2. $f + 17 = 31; f = 14$

3. $108 = 6d; d = 13$

4. $s \div 15 = 9; s = 105$

5. $k \div 21 = 8; k = 168$

6. $45 - p = 18; p = 27$

7. $h \times 17 = 119; h = 7$

8. $73 + m = 24; m = 49$

Identify the solution to each equation from the list given.

9. $b = 16 + 19; 35, 36, 37$

10. $j = 64 - 46; 8, 18, 110$

11. $12n = 168; 14, 15, 16$

12. $56 + w = 81; 15, 25, 35$

13. $m \div 4 = 23; 92, 93, 94$

14. $114 = 6c; 17, 18, 19$

15. $77 - g = 18; 49, 59, 69$

16. $18 = 126 \div y; 5, 6, 7$

Solve each equation mentally.

17. $8x = 56$

18. $12 - q = 7$

19. $17 = 8 + z$

20. $7 = 21 \div e$

21. $12 \div 6 = u$

22. $72 = 9t$

23. $r = 14 - 8$

24. $y + 9 = 13$

25. $9 = n \div 6$

26. $15 - h = 6$

27. $7s = 49$

28. $p \div 2 = 12$

2-1 Study Guide

Frequency Tables

Members of a sixth-grade class were surveyed to determine when to hold the annual trip to the amusement park. The results are shown at the right. On what date should the trip be held?

Dates Reported			
June 9	June 2	June 9	June 8
June 2	June 9	June 8	June 9
June 9	June 2	June 9	June 1
June 2	June 9	June 9	June 9
June 8	June 2	June 9	June 1
June 2	June 9	June 2	June 2
June 8	June 9		

Explore What do you know?
You know how each person responded.

What do you need to know?
You are deciding on what date to take the trip.

Plan Make a frequency table.

Solve The frequency table shows that the greatest number of people want to take the trip on June 9.

Date	Tally	Frequency
June 1	II	2
June 2	HHT III	8
June 8	IIII	4
June 9	HHT HHT II	12

Examine Since June 9 received the greatest number of votes, the annual trip should be held on June 9.

Solve.

1. The ages of students in the drama club at Grindstone Middle School are shown below.

11	12	11	13	14	13	13
12	12	12	12	11	10	13
13	14	12	11	12	13	

   a. Make a frequency table for the data.

   b. What was the most common age of the drama club members?

2. Twenty-one people were asked to name the state in which they were born. The data are shown below.

NY	TX	PA	NY	NY	TX	CA
NY	CA	CA	CA	NY	PA	PA
NY	NY	PA	NY	NY	CA	TX

   a. Make a frequency table for the data.

   b. In which state were the most people surveyed born?

   c. How many of the people surveyed were born in New York State?

2-1 Practice

Frequency Tables

Make a frequency table for each set of data.

1. zip codes of people who participated in a telephone survey.

43081	43231	43223	43229	43227	43082	43017	43229
43085	43081	43229	43231	43082	43085	43229	43231
43082	43085	43085	43017	43229	43227	43017	43017

2. favorite ice creams of sixth-graders at Blendon Middle School: butter pecan (BP), chocolate (C), rocky road (RR), strawberry (S), vanilla (V)

V	C	BP	V	V	V	V	RR
BP	BP	C	S	V	C	S	BP
RR	BP	C	S	V	V	C	C
C	RR	BP	BP	V	S	C	RR

3. number of pets given by people for a pet store survey

3	2	3	3	0	1	1	4
2	2	2	3	4	1	2	1
5	1	1	1	3	2	5	2
1	1	2	2	2	3	3	2
4	1	0	1	3	2	4	

4. ages of the members of the Smith-Jones bridal party

25	27	24	23	6	5	30	22
16	18	29	32	25	22	18	27

5. population (in millions) of the 10 most-heavily populated cities in the United States

Chicago, 3	Dallas, 1	Detroit, 1	Houston, 2	Los Angeles, 3
New York, 7	Philadelphia, 2	Phoenix, 1	San Antonio, 1	San Diego, 1

6. amounts of money found in the pockets of slacks at Clean-Dry over the course of one year

$36	$5	$1	$20	$47	$51	$20	$10
$10	$5	$5	$5	$1	$20	$36	$1

7. favorite pizza toppings of members of the football team at Drew High: extra cheese (C), mushrooms (M), pepperoni (P), sausage (S), vegetarian (V)

P	P	S	S	V	P	V	C
M	M	V	P	S	C	C	M

Mathematics: Applications and Connections, Course 1

2-2 Study Guide

Scales and Intervals

The employees of Lake Products Corporation earn the following yearly salaries: $14,500; $26,000; $43,200; $23,700; $33,400; $15,500; $28,900; $31,100; $56,300; $41,000; $35,000; $24,700; $16,300; $20,000; $63,000; $8,100; $22,800; $9,700; $32,200; $19,300.

A scale from $0 to $69,999 may be used to make a frequency table that includes all of the salaries. The salaries may be divided into seven sections using an interval of $9,999.

The frequency table shows that most employees make between $10,000 and $39,999.

Employee Salaries		
Salary	Tally	Frequency
$60,000–$69,999	I	1
$50,000–$59,999	I	1
$40,000–$49,999	II	2
$30,000–$39,999	IIII	4
$20,000–$29,999	HHI	5
$10,000–$19,999	HHI	5
$0–$9,999	II	2

Determine the scale for a frequency table for each set of data.

1. 63, 89, 12, 45, 83, 62, 91, 44, 62, 42, 77, 9

2. 3,482; 1,955; 8,312; 6,605; 850; 7,831; 300; 5,067

3. 702, 593, 769, 104, 333, 210, 678, 996, 384, 339, 652, 151

Choose the best interval for a frequency table for each set of data.

4. the data in Exercise 1
   a. 1   b. 10   c. 100

5. the data in Exercise 2
   a. 10   b. 100   c. 1,000

6. the data in Exercise 3
   a. 10   b. 100   c. 1,000

Choose the best number to end a scale for a frequency table for each set of data.

7. the data in Exercise 1
   a. 0   b. 10   c. 100

8. the data in Exercise 2
   a. 300   b. 1,000   c. 9,000

9. the data in Exercise 3
   a. 100   b. 1,000   c. 9,000

Scales and Intervals

Choose the better scale for a frequency table for each set of data.

1. 27, 15, 39, 21, 58, 34, 46   **a.** 0 to 60   **b.** 10 to 50

2. 415, 258, 109, 374, 583, 116, 497, 601, 562   **a.** 0 to 600   **b.** 0 to 1,000

3. 18, 25, 40, 27, 11, 42, 13, 9, 36, 5   **a.** 0 to 50   **b.** 0 to 100

4. 5,096, 3,237, 8,924, 2,143, 9,451, 3,670, 7,182   **a.** 0 to 10,000   **b.** 2,000 to 10,000

5. 3, 4, 8, 6, 5, 3, 9, 7, 5, 6   **a.** 0 to 10   **b.** 1 to 5

6. 206, 247, 268, 219, 232, 193, 271, 254, 285   **a.** 0 to 300   **b.** 150 to 300

7. 13,500, 26,700, 39,200, 61,300, 42,800, 11,900, 25,600   **a.** 0 to 70,000   **b.** 10,000 to 70,000

8. $96, $54, $78, $92, $63, $41, $59, $85, $97, $60   **a.** $50 to $90   **b.** $40 to $100

Choose the best interval for a frequency table for each set of data.

9. the data in Exercise 1
    a. 1   **b.** 10   **c.** 100

10. the data in Exercise 2
    a. 1   **b.** 10   **c.** 100

11. the data in Exercise 3
    a. 1   **b.** 5   **c.** 100

12. the data in Exercise 4
    a. 1,000   **b.** 5,000   **c.** 10,000

13. the data in Exercise 5
    a. 1   **b.** 10   **c.** 100

14. the data in Exercise 6
    a. 1   **b.** 15   **c.** 150

15. the data in Exercise 7
    a. 100   **b.** 1,000   **c.** 10,000

16. the data in Exercise 8
    a. 1   **b.** 10   **c.** 100

Make a frequency table for each set of data.

17. the data in Exercise 1

18. the data in Exercise 2

19. the data in Exercise 3

20. the data in Exercise 4

21. the data in Exercise 5

22. the data in Exercise 6

2-3 Study Guide

Bar Graphs and Line Graphs

The diagram shows the parts of a graph.

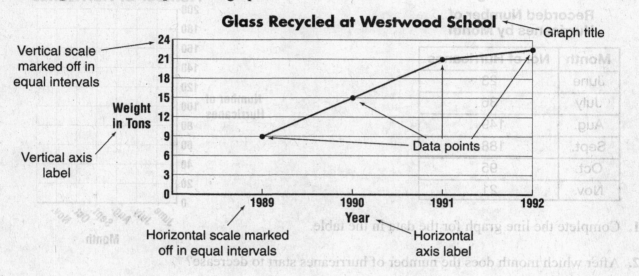

Vertical scale marked off in equal intervals

Weight in Tons

Vertical axis label

Glass Recycled at Westwood School — Graph title

Data points

Horizontal scale marked off in equal intervals

Horizontal axis label

Solve.

1. Make a bar graph for this set of data.

Class President Election Results	
Name	**Number of Votes**
Joyce	18
Ron	11
Ramona	15
Chi Wan	9

2. Make a line graph for this set of data.

Evans Family Electric Bill	
Month	**Amount**
March	$129.90
April	$112.20
May	$105.00
June	$88.50

2-3 Practice

Bar Graphs and Line Graphs

Refer to the following table for Exercises 1-2.

Recorded Number of Hurricanes by Month

Month	No. of Hurricanes
June	23
July	36
Aug.	149
Sept.	188
Oct.	95
Nov.	21

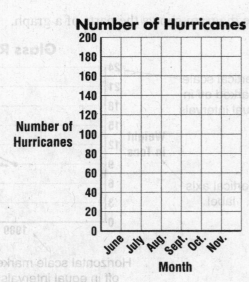

Number of Hurricanes

1. Complete the line graph for the data in the table.

2. After which month does the number of hurricanes start to decrease?

3. Use the data in the table below to complete the bar graph.

Favorite Ice Cream Flavors

Flavor	Frequency
Chocolate	36
Vanilla	28
Strawberry	14
Coffee	15
Maple Walnut	7

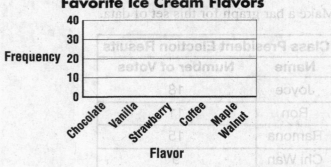

Favorite Ice Cream Flavors

Solve. Use the bar graph.

4. Which city has the highest average number of inches of precipitation?

5. What is the approximate average number of inches of precipitation for Washington D.C.?

6. What is the difference between the average number of inches of precipitation between San Francisco and Salt Lake City?

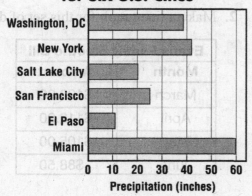

Average Precipitation in Inches for Six U.S. Cities

2-4 **Study Guide**

Reading Circle Graphs

Circle graphs show parts of a whole.

For this graph, the whole is all of the
dog food sold. The parts are the types of
dog food.

How much more of the dog food sold is
dry than canned?

To solve, subtract the canned part from
the dry part.

$$\begin{array}{r} 70\% \\ -27\% \\ \hline 43\% \end{array}$$

43% more of the dog food sold is dry.

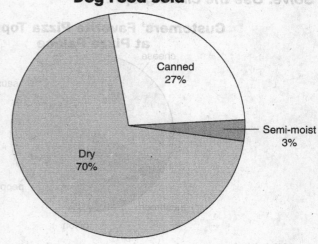

Dog Food Sold

Canned 27%
Semi-moist 3%
Dry 70%

Solve. Use the circle graph.

1. What is the greatest amount of oil
   used for?

2. What is the least amount of oil used
   for?

3. How much of the oil is used
   for heating and for electricity
   generation?

4. How much more oil is used for
   transportation than is used in
   industry?

5. How much greater is the amount of
   oil used for transportation than the
   amount of oil used for other
   purposes?

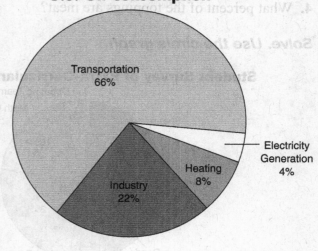

U.S. Oil Consumption

Transportation 66%
Electricity Generation 4%
Heating 8%
Industry 22%

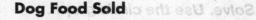

2-4 Practice

Reading Circle Graphs

Solve. Use the circle graph.

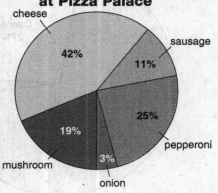

Customers' Favorite Pizza Toppings at Pizza Palace

cheese 42%
sausage 11%
pepperoni 25%
onion 3%
mushroom 19%

1. Which pizza topping is the most popular?

2. Which pizza topping is the least popular?

3. Which two toppings together are about as popular as cheese?

4. What percent of the toppings are meat?

Solve. Use the circle graph.

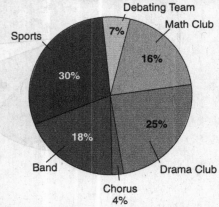

Student Survey of Extra-Curricular Activities

Debating Team 7%
Math Club 16%
Sports 30%
Drama Club 25%
Band 18%
Chorus 4%

5. Which extracurricular activity is the least popular?

6. Which extracurricular activity is about as popular as Band?

7. Which two activities together are about as popular as Drama Club?

2-5 Study Guide

Making Predictions

With this **line graph,** you can make predictions about the expected number
of calories used while bicycling or playing tennis.

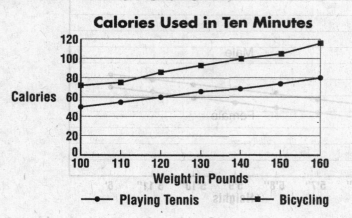

Calories Used in Ten Minutes

Weight in Pounds

● Playing Tennis   ■ Bicycling

Example **About how many more calories can a 120-pound person
expect to use bicycling for 10 minutes than playing
tennis for 10 minutes?**

From the graph, a 120-pound person will use about 60
calories playing tennis and about 82 calories bicycling for
10 minutes.

82 − 60 = 22

A 120-pound person can expect to use about 22 more
calories in 10 minutes of bicycling.

Refer to the graph.

1. About how many calories can a 150-pound
person expect to use bicycling for 10
minutes?

2. About how many calories can a 110-
pound person expect to use playing
tennis for 10 minutes?

3. About how many more calories can a 140-
pound person expect to use bicycling for 10
minutes than playing tennis for 10 minutes?

4. About how many more calories will a
160-pound person use playing tennis for
10 minutes than a 100-pound person will
use?

2-5 Practice

Making Predictions

Refer to the graph for Exercises 1-6.

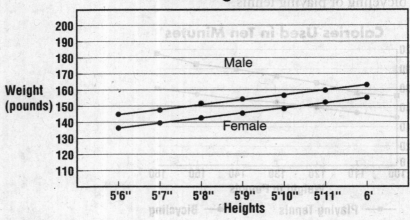

1. About how much should a woman weigh if she is 5'8"?

2. About how much should a man weigh if he is 5'11"?

3. About how tall should a man be if his ideal weight is 155 pounds?

4. About how tall should a woman be if her ideal weight is 140 pounds?

5. What would you expect the approximate ideal weight to be for a man 6'3" tall?

6. What would you expect the approximate ideal weight to be for a woman 6'1" tall?

Refer to the temperature graph for Exercises 7-9.

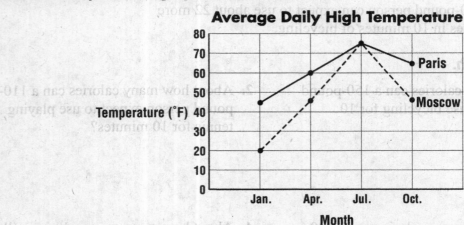

7. Give the expected high temperature for Moscow in October.

8. Give the expected high temperature for Paris in April.

9. What month is the average daily high temperature the same for both Moscow and Paris?

2-6 Study Guide

Stem-and-Leaf Plots

The chart below shows the number of home runs Hank Aaron hit by year.

Home Runs Hit by Hank Aaron											
Year	1954	1955	1956	1957	1958	1959	1960	1961	1962	1963	1964	1965
Home Runs	13	27	26	44	30	39	40	35	45	44	24	32
Year	1966	1967	1968	1969	1970	1971	1972	1973	1974	1975	1976	
Home Runs	44	39	29	44	38	47	34	40	20	12	10	

A **stem-and-leaf plot** of the data is shown below.

The tens digits are the stems.
The ones digits are the leaves.
They are arranged in each row from least to greatest.
A key is included.

Stem	Leaf
1	0 2 3
2	0 4 6 7 9
3	0 2 4 5 8 9 9
4	0 0 4 4 4 4 5 7

4 | 0 means 40.

Make a stem-and-leaf plot for each set of data.

1. 89, 54, 67, 78, 65, 89, 57, 87, 75, 59, 65, 72, 60, 73, 65

2. 85, 124, 90, 113, 107, 94, 88, 114, 106, 109, 110, 117, 100, 101, 119

Use the stem-and-leaf plot for Hank Aaron's home runs.

3. What was the greatest number of home runs in a year? What was the fewest home runs in a year?

4. How many years did Hank Aaron hit 30 or more home runs?

2-6 Practice

Stem-and-Leaf Plots

Determine the stems for each set of data.

1. 44, 32, 77, 44, 31, 45, 79, 34, 35, 66, 55

2. 56, 48, 90, 69, 82, 91, 44, 55, 60, 72

3. 5, 3, 33, 58, 22, 39, 40, 38, 22, 57, 29

4. 20, 13, 15, 6, 16, 29, 24, 22, 21, 20

5. 89, 134, 79, 65, 85, 132, 101, 88, 100

6. 94, 68, 90, 35, 84, 92, 103, 88, 91, 80

Make a stem-and-leaf plot for each set of data.

7. 18, 67, 35, 20, 45, 55, 69, 23, 34, 58, 61, 43, 56, 63, 29, 32

8. 82, 91, 80, 105, 113, 104, 83, 90, 84, 91, 109, 112, 100, 92, 85, 92, 92

9. $1.13, $1.25, $1.19, $1.32, $1.25, $1.50, $1.45, $1.48, $1.52, $1.19

10. $0.89, $1.12, $0.92, $1.28, $1.25, $1.02, $1.13, $1.02, $1.01, $1.10, $1.14, $1.23

Use the stem-and-leaf plot to answer the following.

11. How many zero-degree days does the coldest metropolitan area in the United States have in a year?

12. What is the range of the zero-degree day data?

Number of zero-degree days per year in the eight coldest metropolitan areas of the United States	
Stem	Leaf
3	1 3 4 5
4	1
5	1 1 4
3|1 = 31 zero-degree days per year	

Mathematics: Applications and Connections, Course 1

2-7 Study Guide

Mean, Median, and Mode

Mean, median and mode are measures of central tendency.

To find the **mean** of a set of numbers, find the sum of the numbers and divide by the number of addends.

To find the **median** of a set of numbers, arrange the numbers in order and find the middle number.

To find the **mode** of a set of numbers, find the number that appears most often.

To find the **range** of a set of numbers, subtract the least number from the greatest number.

Student Heights in Inches		
65	62	66
59	62	60
64	59	66
62	64	67

Example   Find the mean, median, and mode of the student heights.

Mean: $\dfrac{65 + 62 + 66 + 59 + 62 + 60 + 64 + 59 + 66 + 62 + 64 + 67}{12} = 63$

Median: 59 59 60 62 62 62 64 64 65 66 66 67

$(62 + 64) \div 2 = 63$

Mode: 62

Range: $67 - 59$ or 8

Find the mean, median, mode, and range for each set of data.

1. 8, 10, 6, 9, 8, 7

2. 12, 6, 8, 2, 7, 5, 2

3. 11, 9, 6, 14, 5, 5, 13

4. 20, 30, 40, 10, 20, 90, 70

5. 16, 20, 18, 14, 17

6. 17, 31, 29, 42, 17, 36, 24

7. 152, 148, 150

2-7 Practice

Mean, Median, and Mode

Find the mean, median, mode, and range for each set of data.

1. 6, 9, 2, 4, 3, 6, 5

2. 25, 18, 14, 27, 25, 14, 18, 25, 23

3. 453, 345, 543, 345, 534

4. 13, 6, 7, 13, 6

5. 8, 2, 9, 4, 6, 8, 5

6. 13, 7, 17, 19, 7, 15, 11, 7

7. 1, 15, 9, 12, 18, 9, 5, 14, 7

8. 28, 32, 23, 43, 32, 27, 21, 34

9. 3, 9, 4, 3, 9, 4, 2, 3, 8

10. 42, 35, 27, 42, 38, 35, 29, 24

11. 157, 124, 157, 124, 157, 139

Use the data below to answer Exercises 12-15.

12. What is the mode?

13. What is the median?

14. What is the mean?

15. If the price of swimming lessons went up to $6,
    would it change the
    a. mode? Why?

    b. median? Why?

    c. mean? Why?

    d. range? Why?

After-School Activities	Cost per Hour
Piano	$14.00
Ballet	10.00
Swimming	5.00
Karate	15.00
Skating	12.00
Aerobics	5.00
Arts & Crafts	8.00
Drama	10.00
Scouts	2.00

Mathematics: Applications and Connections, Course 1

2-8 Study Guide

Misleading Statistics

Some graphs are misleading. Watch for scales with unequal intervals and missing titles and labels on scales.

Example

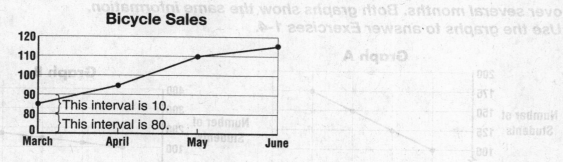

Bicycle Sales

This interval is 10.
This interval is 80.

Use the best measure of central tendency.

Measure of Central Tendency	When to Use
Mean	when no numbers are much greater or much less than the others
Median	when there are numbers that are much greater or much less than the others
Mode	when the most frequently occuring number is needed

Use the chart and a measure of central tendency to show each of the following.

1. weight

2. age

3. hair color

Name	Bill	Joe	Yi	Tony	Al
Age	43	39	42	45	12
Weight	162	155	168	182	148
Hair	brown	blond	black	blond	blond

Solve. Use the line graph above.

4. Draw a scale that could be used to make a graph that is not misleading.

5. Change the title of the graph so that it is misleading.

Mathematics: Applications and Connections, Course 1

Practice

Misleading Statistics

A school lunchroom aide made two graphs to show the number of students participating in the school lunch program over several months. Both graphs show the same information. Use the graphs to answer Exercises 1-4.

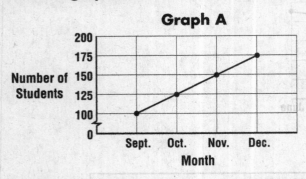

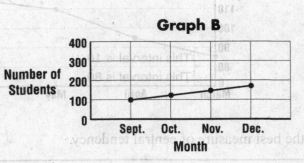

1. Which graph makes the school lunch program look less increasingly popular with students?

2. Which graph would suggest that some new items should be added to the lunch menu? Why?

3. Which graph would you show to the head of the entire school lunch program?

4. What is misleading about Graph B?

Tell whether the mean, median, or mode would be best to describe the following. Explain each answer.

5. the most popular candidate for student council president

6. the results of an easy math quiz in an honors class

7. 85, 92, 87, 93, 90, 88

8. 125, 406, 110, 537, 118, 129, 113, 124

9. average yearly snowfall for the six New England States

2-9 Study Guide

Integration: Geometry
Graphing Ordered Pairs

In mathematics, we locate points by using a **coordinate system.** This system is formed when two number lines intersect at their zero points. This point is called the **origin,** and is labeled with an O. The horizontal number line is called the **x-axis,** and the vertical number line is called the **y-axis.**

You can name any point on a coordinate system by using an **ordered pair** of numbers. The first number is called the **x-coordinate,** and the second number is called the **y-coordinate.**

Examples

1 **Name the ordered pair for point W.**

Start at the origin. Move right along the x-axis until you are under point W. Since you moved three units to the right, the x-coordinate of the ordered pair is 3.

Now move up until you reach point W. Since you moved up 1 unit, the y-coordinate is 1. The ordered pair for point W is (3, 1).

2 **Graph $P(0, 2)$.**

Start at the origin. Since the x-coordinate is 0, do not move any units to the right on the x-axis. Move 2 units up to locate the point. Draw a dot and label the point P.

Use the grid at the right to name the point for each ordered pair.

1. (0, 4)   2. (3, 6)

3. (2, 1)   4. (1, 8)

5. (7, 5)   6. (5, 3)

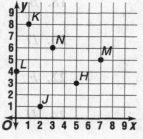

Graph each point.

7. $A(1, 5)$ 8. $F(9, 2)$

9. $X(6, 0)$ 10. $D(4, 8)$

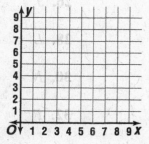

2-9 Practice

Integration: Geometry
Graphing Ordered Pairs

Use the grid at the right to name the point for each ordered pair.

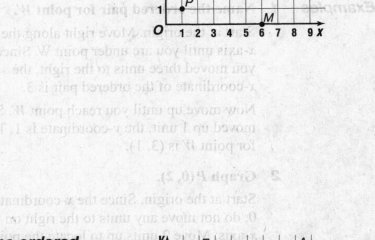

1. (1, 9)
2. (4, 5)
3. (1, 4)
4. (6, 9)
5. (8, 9)
6. (7, 4)
7. (4, 2)
8. (4, 8)
9. (0, 7)
10. (6, 2)
11. (9, 2)
12. (7, 7)
13. (2, 6)
14. (1, 1)
15. (6, 0)
16. (9, 6)

Use the grid at the right to name the ordered pair for each point.

17. A
18. B
19. C
20. D
21. E
22. F
23. G
24. H
25. I
26. J
27. K
28. L
29. M
30. N
31. O
32. P

16

Mathematics: Applications and Connections, Course 3

3-1 Study Guide

Decimals Through Ten-Thousandths

Fraction: $\frac{5,016}{10,000}$　　　　Decimal: 0.5016

Say: five thousand sixteen ten-thousandths

Ones	Tenths	Hundredths	Thousandths	Ten-thousandths
0 •	5	0	1	6

Here are some other examples.

Fraction	Decimal	Words
$\frac{924}{1,000}$	0.924	Nine hundred twenty-four thousandths
$5\frac{7}{10}$	5.7	Five and seven tenths

Write each fraction as a decimal.

1. $\frac{31}{100}$　　　　2. $\frac{9}{100}$　　　　3. $\frac{4}{10,000}$　　　　4. $\frac{35}{1,000}$

5. $\frac{1,654}{10,000}$　　　　6. $\frac{1}{10}$　　　　7. $\frac{6}{1,000}$　　　　8. $\frac{3}{10}$

Write each expression as a decimal.

9. two hundred fifty-one thousandths

10. one and eleven hundredths

11. eight hundredths

12. seventy and fifty-six thousandths

13. five hundred and two ten-thousandths

14. thirty-six ten-thousandths

3-1 Practice

Decimals Through Ten-Thousandths

Write each fraction or mixed number as a decimal.

1. $\frac{8}{10}$ 2. $\frac{19}{100}$ 3. $\frac{7}{100}$ 4. $\frac{26}{100}$

5. $\frac{32}{100}$ 6. $\frac{5}{100}$ 7. $\frac{4}{10}$ 8. $\frac{11}{100}$

9. $\frac{6}{100}$ 10. $\frac{48}{100}$ 11. $\frac{93}{100}$ 12. $\frac{2}{10}$

13. $\frac{407}{1,000}$ 14. $\frac{9}{1,000}$ 15. $\frac{2,351}{10,000}$

16. $\frac{63}{10,000}$ 17. $\frac{742}{1,000}$ 18. $\frac{8}{10,000}$

19. $\frac{914}{10,000}$ 20. $\frac{3,806}{10,000}$ 21. $\frac{59}{1,000}$

Write each expression as a decimal.

22. thirteen hundredths

23. two and forty-nine hundredths

24. six and eight hundredths

25. thirty-nine and two tenths

26. eighty-three hundredths

27. seven tenths

28. forty-five and two ten-thousandths

29. thirty-one thousandths

30. four thousandths

31. twelve and nine hundred five ten-thousandths

Mathematics: Applications and Connections, Course 1

3-2

Study Guide

Integration: Measurement
Length in the Metric System

A dime is about one millimeter
(1 mm) thick.

1 mm

A sugar cube is about one centimeter
(1 cm) wide.

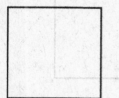

1 cm

A kitchen counter is about one
meter (1 m) high.

Nine times the length of a football
field, including end zones, is about
1 kilometer (1 km).

Use a centimeter ruler to measure each line segment.

1. ▬▬▬   2. ▬▬▬▬▬▬▬▬▬▬▬▬▬▬

3. ▬▬▬   4. ▬▬▬▬▬▬▬▬▬▬▬▬

5. ▬   6. ▬▬▬▬▬▬▬

Use a centimeter ruler to measure one side of each square.

7.   8. □   9. □

10. □   11. □   12. □

Name_____ Date_____

3-2 Practice

Integration: Measurement
Length in the Metric System

Use a centimeter ruler to measure each line segment.

1. _____

2. _____

3. _____

4. _____

5. _____

6. _____

7. _____

8. _____

9. _____

Use a centimeter ruler to measure one side of each figure.

10.

11.

12.
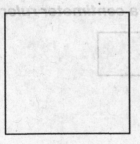

13.

14.
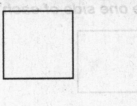
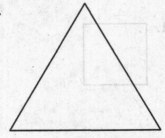

15.
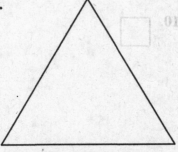

*Mathematics: Applications
and Connections, Course 1*

Name _____ **Date** _____

Study Guide

Comparing and Ordering Decimals

Which is greater, 25.692 or 25.6902?

Line up the decimal points.
Start at the left. Find the
first place in which the
digits are different.

		The decimal with the greater digit is greater.

Compare the digits.

| 25.69**2** | 2 is greater than 0. | 25.692 > 25.6902 |
| 25.69**0**2 | 2 > 0 | |

Order decimals by comparing them two at a time.

State the greater number in each group.

1. 0.042 or 0.422 **2.** 7.398 or 7.378 **3.** 76.1423 or 76.142

4. 1.53 or 1.0053 **5.** 14.358 or 14.374 **6.** 0.092 or 0.192

7. 709.12 or 790.21 **8.** 5.0045 or 5.0405 **9.** 12.2 or 2.222

Order each set of decimals from least to greatest.

10. 6.583	**11.** 456.73	**12.** 0.004
6.843	465.32	0.04
6.065	456.37	0.035
6.269	456.23	0.305

Order each set of decimals from greatest to least.

13. 1.16	**14.** 91.5	**15.** 745.0003
0.616	95.155	745.303
1.066	9.05	745.03
0.06	19.005	745.333
0.016	95.51	745.003

Mathematics: Applications and Connections, Course 1

3-3

Practice

Comparing and Ordering Decimals

Circle the greater number in each group.

1. 6.23 or 6.32

2. 4.58 or 4.5

3. 7.09 or 7.1

4. 0.347 or 0.437

5. 0.92 or 0.095

6. 15.6 or 1.506

7. 28.003 or 28.03

8. 1.406 or 1.064

9. 19.08 or 19.079

10. 0.0705 or 0.075

11. 32.61 or 3.621

12. 40.0598 or 40.589

13. 93.045 or 93.054

14. 21.967 or 2.1968

15. 87.0024 or 87.003

16. 0.849 or 0.0851

17. 6.2709 or 6.2907

18. 5.862 or 52.68

Order each set of decimals from least to greatest.

19. 1.25
    1.52
    1.02
    1.50

20. 18.7
    19.06
    17.9
    18.08

21. 308.629
    380.269
    308.962
    308.296

22. 0.0475
    0.5074
    0.0547
    0.4057

23. 67.39
    68.004
    67.039
    67.04

24. 432.95
    234.96
    342.95
    243.97

Order each set of decimals from greatest to least.

25. 0.60
    0.006
    0.59
    0.059
    0.95

26. 72.052
    71.98
    73.7
    71.89
    72.46

27. 290.41
    289.4
    295.014
    289.14
    295.104

Mathematics: Applications and Connections, Course 1

3-4 Study Guide

Rounding Decimals

Round 34.725 to the nearest tenth.

You can use a number line.

Find the approximate location of   34.725 is closer to 34.7 than to 34.8
34.725 on the number line.   34.725 rounded to the nearest tenth is 34.7.

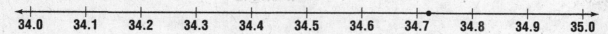

You can also round without a number line.

Find the place to which you want to round.	Look at the digit to the right. If the digit is less than 5, round down. If the digit is 5 or greater, round up.	2 is less than 5. Round down.
34.7<u>2</u>5	34.7<u>2</u>5	34.7

Use each number line to show how to round the decimal to the nearest tenth.

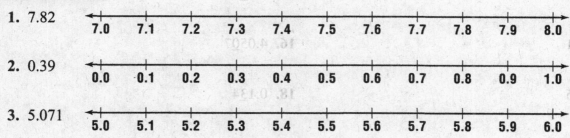

1. 7.82
2. 0.39
3. 5.071

Round each number to the underlined place-value position.

4. 6.3<u>2</u>

5. 0.4<u>7</u>21

6. 26.<u>4</u>44

7. 1.1<u>6</u>1

8. 362.08<u>4</u>6

9. 15.5<u>5</u>3

10. 151.3<u>9</u>1

11. 0.<u>5</u>5

12. 631.00<u>0</u>8

13. 17.3<u>2</u>7

14. 3.<u>0</u>9

15. 1.<u>5</u>8

3-4 Practice

Rounding Decimals

Round each number to the underlined place-value position.

1. 1.7<u>2</u>6

2. 5<u>4</u>.38

3. 0.<u>5</u>8

4. 0.9<u>1</u>42

5. 80.<u>6</u>59

6. 23<u>2</u>.1

7. 1.<u>0</u>63

8. 0.5<u>5</u>

9. 0.<u>8</u>194

10. 0.4<u>9</u>6

11. 3.01<u>8</u>2

12. 71.<u>4</u>05

13. <u>9</u>.63

14. 32.<u>7</u>1

15. 2.6<u>7</u>1

16. 4.05<u>0</u>7

17. 89.<u>9</u>5

18. 0.1<u>3</u>4

19. 5.<u>8</u>93

20. 52<u>0</u>.6

21. 0.70<u>9</u>8

22. 1.8<u>4</u>5

23. 34.<u>5</u>5

24. 29.<u>2</u>5

25. 56.09<u>2</u>4

26. 1,19<u>9</u>.7

27. 0.<u>4</u>6

28. 0.<u>3</u>546

3-5 Study Guide

Estimating Sums and Differences

Two estimation strategies are **rounding** and **clustering**.

Round each number to the same place.

Examples

Round to the nearest ten dollars.	Round to the nearest tenth.	Round to the nearest one.

$$\begin{array}{r}\$46.90 \\ +\ 33.27\end{array} \longrightarrow \begin{array}{r}\$50 \\ +\ 30 \\ \hline \$80\end{array}$$ $$\begin{array}{r}0.693 \\ -\ 0.113\end{array} \longrightarrow \begin{array}{r}0.7 \\ -\ 0.1 \\ \hline 0.6\end{array}$$ $$\begin{array}{r}6.22 \\ +\ 0.85\end{array} \longrightarrow \begin{array}{r}6 \\ +\ 1 \\ \hline 7\end{array}$$

To use clustering, find a number around which each number in the set "clusters." Then use that number for all of the numbers.

Example   **$7.62 + $7.89 + $8.01 + $7.99**   *These numbers cluster around 8.*

Add $8 four times.

$8 + $8 + $8 + $8 = $32

Estimate using rounding.

1. $$\begin{array}{r}0.456 \\ +\ 0.375\end{array}$$

2. $$\begin{array}{r}59.118 \\ -\ 17.799\end{array}$$

3. $$\begin{array}{r}\$6.63 \\ +\ 9.29\end{array}$$

4. $$\begin{array}{r}0.0056 \\ -\ 0.0028\end{array}$$

5. $8.802 - 6.115$

6. $0.9 - 0.0984$

Estimate using clustering.

7. $13.1 + 12.97 + 12.62 + 13.44$

8. $1.01 + 0.67 + 1.39$

9. $$19.99 + $20.15 + 19.52

10. $5.55 + 6.01 + 5.7 + 6.412$

3-5 Practice

Estimating Sums and Differences

Estimate using rounding.

1. 5.62
    +3.04

2. 18.93
    +27.45

3. 9.24
    −2.56

4. 16.72
    −9.13

5. 0.417
    +0.869

6. 42.905
    +31.276

7. 0.754
    −0.482

8. 87.146
    −24.953

9. 0.69
    +0.45

10. 0.74
    −0.18

11. 12.394
    −4.601

12. 24.537
    +9.862

13. 6.521 + 4.378

14. 0.932 − 0.485

15. 0.86 + 0.95

16. 43.058 − 15.726

Estimate using clustering.

17. 59.62 + 60.4 + 60 + 61

18. 8.2 + 7.8 + 7.2 + 7.99

19. 26.08 + 25.99 + 26.17

20. 6.73 + 7.01 + 7.53 + 6.91 + 7.1

21. 3.42 + 3.11 + 2.9 + 2.6

22. 4.59 + 5.28 + 5.444

23. 67.24 + 66.905 + 65 + 67.3

24. 87.04 + 86.55 + 87.101 + 86

25. $4.79 + $5.29 + $4.99

26. 9.634 + 9.9 + 9.46 + 9.91 + 9.7632

27. 3.604 + 3.918 + 3.342 + 4.1

28. 2.1 + 2.387 + 2.57 + 1.99

3-6 Study Guide

Adding and Subtracting Decimals

To add decimals, line up the decimal points.
Then add the same way you add whole numbers.

Examples **1** Find 0.465 + 0.292.

$$\begin{array}{r} 0.465 \\ + \ 0.292 \\ \hline 0.757 \end{array}$$

2 Find 13.2 + 5.08 + 0.334.

$$\begin{array}{r} 13.200 \\ 5.080 \\ + \ 0.334 \\ \hline 18.614 \end{array}$$ ←—Annex zeros if necessary.

To subtract decimals, line up the decimal points. Then subtract
the same way you subtract whole numbers.

Examples **3** Find $155.36 − $24.17.

$$\begin{array}{r} \$155.36 \\ - \ 24.17 \\ \hline \$131.19 \end{array}$$

4 Find 0.56 − 0.0277.

$$\begin{array}{r} 0.5600 \\ - \ 0.0277 \\ \hline 0.5323 \end{array}$$ ←—Annex zeros if necessary.

Add or subtract.

1. $\begin{array}{r} 0.352 \\ + \ 0.365 \end{array}$

2. $\begin{array}{r} 78.158 \\ - \ 17.326 \end{array}$

3. $\begin{array}{r} \$9.63 \\ + \ 7.29 \end{array}$

4. $\begin{array}{r} 0.0123 \\ - \ 0.0028 \end{array}$

5. $\begin{array}{r} 2.011 \\ + \ 5.852 \end{array}$

6. $\begin{array}{r} \$13.67 \\ - \ 7.19 \end{array}$

7. $\begin{array}{r} 0.0783 \\ + \ 0.0238 \end{array}$

8. $\begin{array}{r} 9.054 \\ - \ 4.038 \end{array}$

9. 5.014 + 12.3 + 0.4

10. 216.8 − 34.055

Solve each equation.

11. $m = 6.2 − 1.9$

12. $3.07 + 1.1 + 0.9 = k$

13. $\$44.15 − \$1.99 = p$

14. $y = 0.075 + 1.2$

15. $v = 6.12 + 0.3 + 15.4$

16. $9 − 4.06 = c$

3-6 Practice

Adding and Subtracting Decimals

Add or subtract.

1.	4.78 +6.25	2.	6.539 +2.817	3.	12.43 −7.65	4.	1.502 −0.638

5.	28.4 +3.7	6.	70.29 −16.57	7.	67.89 +75.04	8.	604.2 −325.7

9.	0.86 +0.38	10.	1.52 −0.85	11.	47.06 −38.27	12.	7.895 +2.417

13.	9.214 −5.618	14.	16.45 +37.82	15.	50.2 −31.9	16.	49.7 +64.8

17. $7 - 2.6$

18. $58.6 + 13.09$

19. $414 - 357.42$

20. $7.08 + 2.607$

21. $56.18 - 24.093$

22. $0.73 + 18.5 + 9.402$

Solve each equation.

23. $8.3 + 7.9 = n$

24. $k = 35.8 + 24.69$

25. $134 - 57.63 = d$

26. $f = 19.4 - 7.86$

27. $0.485 + 9.32 = g$

28. $j = 362 - 145.9$

29. Evaluate the expression $y - z$ if $y = 40.63$ and $z = 17.85$.

30. Evaluate the expression $j + k$ if $j = 24.25$ and $k = 7.491$.

4-1 Study Guide

Multiplying Decimals by Whole Numbers

When you multiply a decimal by a whole number, multiply as with whole numbers. The product must have the same number of decimal places as the decimal factor.

Examples **1** **Find 6×5.43.**

$$\begin{array}{r} 5.43 \\ \times\ \ 6 \\ \hline 32.58 \end{array}$$ ← two decimal places
← two decimal places

2 **Find 120×0.056.**

$$\begin{array}{r} 0.056 \\ \times\ \ 120 \\ \hline 1\ 120 \\ 5\ 6\ \ \ \\ \hline 6.720 \end{array}$$ ← three decimal places
← three decimal places

Multiply.

1. $\begin{array}{r} 0.7 \\ \times\ \ 9 \\ \hline \end{array}$

2. $\begin{array}{r} 0.78 \\ \times\ \ 17 \\ \hline \end{array}$

3. $\begin{array}{r} 0.09 \\ \times\ \ 101 \\ \hline \end{array}$

4. $\begin{array}{r} 6.2 \\ \times\ \ 12 \\ \hline \end{array}$

5. $\begin{array}{r} 4.12 \\ \times\ \ 22 \\ \hline \end{array}$

6. $\begin{array}{r} 10.4 \\ \times\ \ 221 \\ \hline \end{array}$

7. $\begin{array}{r} 131.5 \\ \times\ \ 55 \\ \hline \end{array}$

8. $\begin{array}{r} 0.3 \\ \times\ \ 494 \\ \hline \end{array}$

9. $3,330 \times 0.05$ 10. 75×0.003 11. 9×5.05

Solve each equation.

12. $f = 8 \times 0.006$ 13. $a = 205 \times 0.22$ 14. $t = 31 \times 1.12$

4-1 Practice

Multiplying Decimals by Whole Numbers

Use estimation to place the decimal point in each product.

1. $0.73 \times 56 = 4088$

2. $2.7 \times 48 = 1296$

3. $2.94 \times 108 = 31752$

4. $1.035 \times 69 = 71415$

5. $0.8 \times 472 = 3776$

6. $15.06 \times 319 = 480414$

Multiply.

7. $\begin{array}{r} 0.9 \\ \times\ 6 \\ \hline \end{array}$

8. $\begin{array}{r} 3.47 \\ \times\ 5 \\ \hline \end{array}$

9. $\begin{array}{r} 0.82 \\ \times\ 9 \\ \hline \end{array}$

10. $\begin{array}{r} 27.3 \\ \times\ 8 \\ \hline \end{array}$

11. $\begin{array}{r} 0.64 \\ \times\ 32 \\ \hline \end{array}$

12. $\begin{array}{r} 5.9 \\ \times\ 174 \\ \hline \end{array}$

13. $\begin{array}{r} 0.0358 \\ \times\ 216 \\ \hline \end{array}$

14. $\begin{array}{r} 4.76 \\ \times\ 95 \\ \hline \end{array}$

15. $\begin{array}{r} 208.7 \\ \times\ 43 \\ \hline \end{array}$

16. $\begin{array}{r} 0.4 \\ \times\ 738 \\ \hline \end{array}$

17. $\begin{array}{r} 1.95 \\ \times\ 4,620 \\ \hline \end{array}$

18. $\begin{array}{r} 0.006 \\ \times\ 87 \\ \hline \end{array}$

19. $\begin{array}{r} 89.2 \\ \times\ 54 \\ \hline \end{array}$

20. $\begin{array}{r} 0.013 \\ \times\ 2,361 \\ \hline \end{array}$

21. $\begin{array}{r} 7.49 \\ \times\ 105 \\ \hline \end{array}$

22. $\begin{array}{r} 2.5 \\ \times\ 3,092 \\ \hline \end{array}$

23. 36×0.07

24. 4.8×235

25. 1.29×614

26. 93×0.57

27. 18×270.9

28. 0.006×315

29. Find the product of 58.2 and 67.

30. What is 1,073 times 2.04?

4-2 Study Guide

Using the Distributive Property

The Distributive Property	
The sum of two addends multiplied by a number is equal to the sum of the products of each addend and the number, $a(b + c) = ab + ac$.	**Example** $5(3 + 4) = 5 \times 3 + 5 \times 4$ $5 \times 7 = 15 + 20$ $35 = 35$

The distributive property allows you to solve problems in parts. This makes it easy to solve some multiplication problems mentally.

Examples   **1**   **Find 6×18 mentally using the distributive property.**

$$6 \times 8 = 6(10 + 8) \qquad \textit{Write 18 as } 10 + 8.$$
$$= 6 \times 10 + 6 \times 8 \qquad \textit{Use the distributive property.}$$
$$= 60 + 48 \qquad \textit{Multiply.}$$
$$= 108 \qquad \textit{Add.}$$

   2   **Find 50×20.3 mentally using the distributive property.**

$$50 \times 20.3 = 50(20 + 0.3) \qquad \textit{Write 20.3 as } 20 + 0.3.$$
$$= 50 \times 20 + 50 \times 0.3 \qquad \textit{Use the distributive property.}$$
$$= 1,000 + 15 \qquad \textit{Multiply.}$$
$$= 1,015 \qquad \textit{Add.}$$

Find each product mentally. Use the distributive property.

1. 8×14

2. 106×6

3. 40×2.1

4. 9×2.4

5. 30.3×5

6. 0.7×12

7. 6.7×60

8. 6.6×4

9. 90×1.02

10. 70×1.7

11. 22×1.1

12. 5.7×2

4-2 Practice

Using the Distributive Property

Rewrite each expression using the distributive property.

1. $4(30 + 6)$

2. $0.5(20 + 9)$

3. $7(8 + 0.4)$

4. $16(50 + 2)$

5. $3(200 + 7)$

6. $2.8(10 + 3)$

7. $90(60 + 0.5)$

8. $4(70 + 0.1)$

9. $11(300 + 20)$

Find each product mentally. Use the distributive property.

10. 26×5

11. 0.7×34

12. 9.8×2

13. 14×3.02

14. 50×6.7

15. 108×12

16. 11×24

17. 3×290

18. 80.4×5

19. 7×63

20. 51×0.9

21. 40.2×30

22. 8×2.7

23. 60×5.4

24. 13×12

25. 90×20.1

26. 7×508

27. 16×105

28. Find the product of 4.09 and 80 mentally.

29. Find the product of 6 and 7.5 mentally.

30. Find the product of 21 and 31 mentally.

4-3 Study Guide

Multiplying Decimals

Multiply decimals just as you multiply whole numbers. The number of decimal places in the product is equal to the sum of the number of decimal places in the two factors.

Example   **Multiply 0.16 and 1.025.**

$$
\begin{array}{r}
1.025 \quad \longleftarrow \textit{ three decimal places} \\
\times\ 0.16 \quad \longleftarrow \textit{ two decimal places} \\
\hline
6150 \\
1025\ \\
\hline
0.16400 \quad \longleftarrow \textit{ five decimal places}
\end{array}
$$

Multiply.

1. 0.5×20.2

2. 1.2×2.3

3. 0.055×3.2

4. 0.014×0.4

5. 12.4×12.4

6. 3.07×1.07

Solve each equation.

7. $c = 15.5 \times 3.3$

8. $x = 202.1 \times 1.14$

9. $a = 0.008 \times 65.3$

Evaluate each expression if m = 0.09, n = 1.2, and p = 8.19.

10. mn

11. $p(n + m)$

12. pm

4-3 Practice

Multiplying Decimals

Use estimation to place the decimal point in each product.

1. $18.6 \cdot 4.2 = 7812$

2. $51.9 \cdot 2.73 = 141687$

3. $6.3 \cdot 0.098 = 06174$

4. $7.05 \cdot 42.01 = 2961705$

5. $13.4 \cdot 9.65 = 129310$

6. $0.72 \cdot 1.408 = 101376$

Multiply.

7. $2.5 \cdot 6.7$

8. $0.4 \cdot 8.3$

9. $0.6 \cdot 0.91$

10. $8.54 \cdot 3.27$

11. $0.2 \cdot 0.079$

12. $16.8 \cdot 4.5$

13. $39.6 \cdot 2.417$

14. $0.003 \cdot 2.9$

15. $5.7 \cdot 18.4$

16. $0.93 \cdot 6.8$

17. $0.0004 \cdot 0.05$

18. $200.41 \cdot 3.06$

19. $9.3 \cdot 0.087$

20. $6.5 \cdot 0.9$

21. $4.07 \cdot 23.8$

22. $3.5 \cdot 19.2$

23. $2.1 \cdot 0.76$

24. $50.4 \cdot 1.032$

Solve each equation.

25. $9.5 \cdot 17.34 = p$

26. $r = 28.6 \cdot 0.007$

27. $5.32 \cdot 104.9 = m$

Evaluate each expression if $x = 4.8$, $y = 12$, and $z = 0.036$.

28. yz

29. $(y - x)z$

30. $x(y + z)$

Mathematics: Applications
and Connections, Course 1

4-4 Study Guide

Integration: Geometry
Perimeter and Area

The **perimeter** *P* of a figure is the distance around the figure. You can find the perimeter by adding the measures of the sides of the figure. The **area** *A* is the number of square units needed to cover a surface.

Figure	rectangle	square
Perimeter	The perimeter of a rectangle equals 2 times the length ℓ plus 2 times the width *w*. $P = 2\ell + 2w$	The perimeter of a square equals 4 times the length of one side *s*. $P = 4s$
Area	The area of a rectangle equals the product of its length ℓ and its width *w*. $A = \ell w$	The area of a square equals the square of the length of one side *s*. $A = s^2$
Examples	$P = 2\ell + 2w$ $P = 2 \times 20 + 2 \times 18$ $P = 40 + 36$ or 76 $A = \ell w$ $A = 20(18)$ or 360 18 ft 20 ft	$P = 4s$ $P = 4(4.5)$ or 18 $A = s^2$ $A = (4.5)^2$ $A = 20.25$ $s = 4.5$ m

Find the perimeter of each figure.

1. 12 in. 9 in. 5 in. 15 in.

2.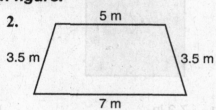
5 m 3.5 m 3.5 m 7 m

3.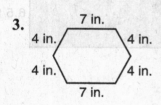
7 in. 4 in. 4 in. 4 in. 4 in. 7 in.

Find the perimeter and area of each figure.

4.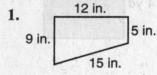
6.8 cm

5.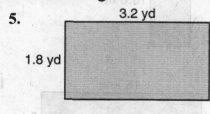
3.2 yd 1.8 yd

6.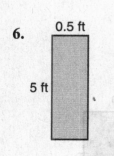
0.5 ft 5 ft

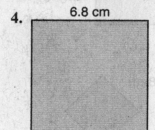

4-4 Practice

4-4 Study Guide

Integration: Geometry
Perimeter and Area

Find the perimeter of each figure.

1.
16.2 m
7.9 m

2.
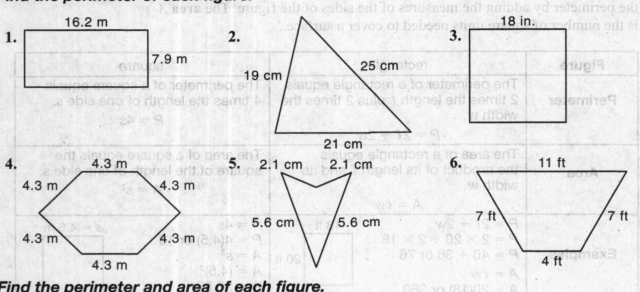
19 cm
25 cm
21 cm

3.
18 in.

4.
4.3 m
4.3 m
4.3 m
4.3 m
4.3 m
4.3 m

5.
2.1 cm   2.1 cm
5.6 cm   5.6 cm

6.
11 ft
7 ft   7 ft
4 ft

Find the perimeter and area of each figure.

7.
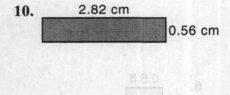
13.2 ft
6.5 ft

8.
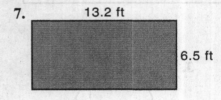
0.98 m

9.
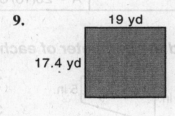
19 yd
17.4 yd

10.
2.82 cm
0.56 cm

11.
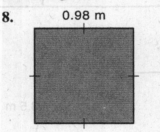
7.3 in.
18.6 in.

12.
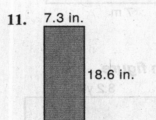
21.7 mm

13.
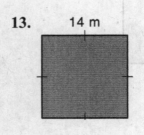
14 m

14.
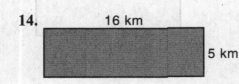
16 km
5 km

15.
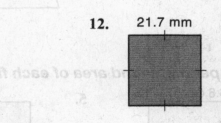
11.8 in.

4-5 Study Guide

Dividing Decimals by Whole Numbers

When you divide a decimal by a whole number, place the decimal point in the quotient above the decimal point in the dividend. Then divide as with whole numbers.

Examples

Place the decimal point above the decimal point in the dividend.

$$12\overline{)38.4}$$

$$8\overline{)6.48}$$

Divide as with whole numbers.

$$\begin{array}{r} 3.2 \\ 12\overline{)38.4} \\ -36 \\ \hline 2\,4 \\ -2\,4 \\ \hline 0 \end{array}$$

$$\begin{array}{r} 0.81 \\ 8\overline{)6.48} \\ -6\,4 \\ \hline 8 \\ -8 \\ \hline 0 \end{array}$$

Find each quotient.

1. $48\overline{)110.4}$

2. $12\overline{)88.8}$

3. $39\overline{)66.3}$

4. $28\overline{)11.76}$

5. $17\overline{)146.2}$

6. $9\overline{)4.68}$

7. $11\overline{)102.3}$

8. $19\overline{)20.9}$

9. $31\overline{)223.2}$

10. $22\overline{)72.6}$

11. $5\overline{)2.95}$

12. $6\overline{)15.6}$

Name_____ Date_____

Dividing Decimals by Whole Numbers

Find each quotient.

1. $7\overline{)57.68}$

2. $32\overline{)14.72}$

3. $15\overline{)58.5}$

4. $26\overline{)13.442}$

5. $49\overline{)308.21}$

6. $78\overline{)452.4}$

7. $59\overline{)273.17}$

8. $17\overline{)3.094}$

9. $4\overline{)3.16}$

10. $92\overline{)625.6}$

11. $5\overline{)7.45}$

12. $48\overline{)12.48}$

13. $15.93 \div 27$

14. $724.12 \div 86$

15. $58.235 \div 95$

16. $25.84 \div 34$

17. $597.8 \div 61$

18. $2.268 \div 3$

Round each quotient to the nearest tenth.

19. $52.4 \div 6$

20. $63.75 \div 34$

21. $948.16 \div 27$

22. $78.29 \div 8$

Round each quotient to the nearest hundredth.

23. $65.24 \div 9$

24. $359.4 \div 75$

25. $58.179 \div 26$

26. $267.54 \div 48$

Mathematics: Applications and Connections, Course 1

4-6 Study Guide

Dividing by Decimals

To divide a decimal by a decimal, first multiply the divisor by a power of ten to make it a whole number. Multiply the dividend by the same power of ten. Then divide as with whole numbers.

Example Find $87.3025 \div 3.715$. *Estimate: $88 \div 4 = 22$*

$$
\begin{array}{r}
23.5 \\
3.715.\overline{)87.302.5} \\
-74\,30 \\
\hline
13\,002 \\
-11\,145 \\
\hline
1\,857\,5 \\
-1\,857\,5 \\
\hline
0
\end{array}
$$

Multiply the divisor and the dividend by 1,000. Place the decimal point in the quotient. Divide.

Find each quotient.

1. $1.4\overline{)9.8}$

2. $2.7\overline{)40.5}$

3. $0.41\overline{)3.69}$

4. $2.1\overline{)4.41}$

5. $0.07\overline{)2.38}$

6. $0.212\overline{)1.696}$

7. $0.013\overline{)0.0208}$

8. $6.28\overline{)87.92}$

Solve each equation.

9. $a = 27.63 \div 0.3$

10. $8.652 \div 1.2 = z$

11. $9.594 \div 0.06 = h$

12. $s = \$1.76 \div 32$

Practice

Study Guide

Dividing by Decimals

Find each quotient.

1. $0.7\overline{)2.52}$ **2.** $3.8\overline{)17.1}$ **3.** $2.64\overline{)150.48}$

4. $5.5\overline{)4.95}$ **5.** $0.09\overline{)0.72}$ **6.** $0.014\overline{)2.184}$

7. $1.32\overline{)3.96}$ **8.** $6.7\overline{)61.64}$ **9.** $0.058\overline{)0.41992}$

10. $34.9\overline{)628.2}$ **11.** $0.48\overline{)308.64}$ **12.** $0.27\overline{)1.593}$

13. $3.6\overline{)2.52}$ **14.** $0.5\overline{)0.105}$ **15.** $0.019\overline{)0.16397}$

16. $0.73\overline{)141.62}$ **17.** $28.6\overline{)42.9}$ **18.** $0.4\overline{)9.52}$

Mathematics: Applications and Connections, Course 1

4-7 Study Guide

Zeros in the Quotient

Remember to write a zero in the quotient when you need a placeholder.

Examples

Use zero as a placeholder in the ones place.

$$
\begin{array}{r}
0.8 \\
7\overline{)5.6} \\
-5\,6 \\
\hline
0
\end{array}
$$

There are no 19s in 2. Use zero as a placeholder.

$$
\begin{array}{r}
0.012 \\
19\overline{)0.228} \\
-\;0 \\
\hline
22 \\
-19 \\
\hline
38 \\
-38 \\
\hline
0
\end{array}
$$

There are no 27s in 18. Use zero as a placeholder.

$$
\begin{array}{r}
1.07 \\
2.7\overline{)2.889} \\
-2\,7 \\
\hline
18 \\
-\;0 \\
\hline
189 \\
-189 \\
\hline
0
\end{array}
$$

Find each quotient to the nearest hundredth.

1. $22\overline{)6.6}$

2. $14\overline{)1.26}$

3. $7\overline{)28.35}$

4. $2.95 \div 59$

5. $0.0264 \div 2.2$

6. $0.158 \div 7.9$

7. $19.065 \div 9.3$

8. $0.102 \div 34$

9. $72.76 \div 6.8$

Practice

Name_____ Date_____

Zeros in the Quotient

Find each quotient to the nearest hundredth.

1. $2.6\overline{)0.208}$

2. $1.2\overline{)0.0648}$

3. $15\overline{)4.53}$

4. $34\overline{)70.04}$

5. $4.2\overline{)42.294}$

6. $56.2\overline{)50.58}$

7. $18\overline{)729}$

8. $6.9\overline{)0.0207}$

9. $17.5\overline{)1.505}$

10. $7.2\overline{)4.68}$

11. $9.5\overline{)288.8}$

12. $3.8\overline{)34.39}$

13. $0.846 \div 1.2$

14. $80.032 \div 1.6$

15. $0.864 \div 24$

16. $2.32 \div 58$

17. $981.4 \div 14$

18. $1.0034 \div 3.46$

19. $0.0378 \div 6.3$

20. $9.652 \div 1.9$

21. $7.828 \div 76$

22. $1,001 \div 25$

23. $2.1843 \div 80.9$

24. $9.66 \div 1.2$

25. $75.84 \div 237$

26. $0.3762 \div 41.8$

27. $5.3 \div 106$

Mathematics: Applications and Connections, Course 1

4-8 Study Guide

Integration: Measurement
Mass and Capacity in the Metric System

Capacity describes the amount a container will hold.

In the metric system, the basic unit of capacity is the **liter** (L).

A 10-centimeter cube has a capacity of 1 L.

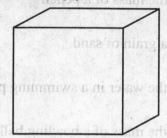

A milliliter (mL) is 0.001 of a liter. 1,000 milliliters equal 1 liter.

A 1-centimeter cube has a capacity of 1 mL.

The quantity of matter an object contains is called **mass.**

In the metric system, the basic unit of mass is the **kilogram** (kg).

A kilogram (kg) is 1,000 grams.

A milligram (mg) is 0.001 gram. 1,000 mg equals 1 g.

A dollar bill has a mass of about 1 gram.

A pair of high top shoes has a mass of about 1 kilogram.

An eyelash has a mass of about 1 milligram.

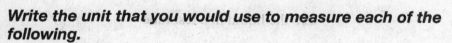

Write the unit that you would use to measure each of the following.

1. a wagon

2. a grain of sugar

3. a swimming pool

4. a horse

5. a peanut

6. a spoonfull of medicine

7. a pail

8. a bag of apples

9. a cherry

10. a bottle of milk

11. a tea pot

12. yourself

4-8 Practice

Study Guide 4-8

Integration: Measurement
Mass and Capacity in the Metric System

Write the unit that you would use to measure each of the following.

1. the mass of a bicycle

2. the mass of a pencil

3. a glass of juice

4. a grain of sand

5. the mass of a penny

6. the water in a swimming pool

7. the mass of a feather

8. the mass of a bowling ball

9. hot chocolate in a large thermos

10. a loaf of bread

11. the mass of a whole watermelon

12. a cup of hot cider

13. the water in an aquarium

14. the mass of a car key

15. the mass of a vitamin pill

16. a can of soup

17. the mass of an egg

18. the mass of a cat

19. the water in a team's cooler

20. the mass of a sewing needle

21. the mass of a mosquito

22. the liquid in a test tube

23. the mass of a student desk

24. the mass of a sandwich

25. a bottle of expensive perfume

26. the mass of a sugar cube

27. the water in a washing machine

28. the mass of a full bag of groceries

29. the mass of an apple

30. the mass of a leaf

4-9 Study Guide

Integration: Measurement
Changing Metric Units

The metric system is a base-10 system. The **meter** is the basic unit of length. The **liter** is the basic unit of capacity. The **kilogram** is the basic unit of mass.

Prefix	Meaning	Length	Capacity	Mass
kilo-	1,000	kilometer (km)	kiloliter (kL)	kilogram (kg)
hecto-	100	hectometer (hm)	hectoliter (hL)	hectogram (hg)
deka-	10	dekameter (dam)	dekaliter (daL)	dekagram (dag)
	1	meter (m)	liter (L)	gram (g)
deci-	0.1	decimeter (dm)	deciliter (dL)	decigram (dg)
centi-	0.01	centimeter (cm)	centiliter (cL)	centigram (cg)
milli-	0.001	millimeter (mm)	milliliter (mL)	milligram (mg)

Units may be changed by multiplying or dividing by multiples of 10.

Examples

1 3,500 mL = _____ L

1,000 mL = 1 L
Divide 3,500 by 1,000.
3,500 mL = 3.5 L

2 7.6 km = _____ m

1km = 1,000 m
Multiply 7.6 by 1,000.
7.6 km = 7,600 m

3 5,000 mg = _____ g

1,000 mg = 1 g
Divide 5,000 by 1,000.
5,000 mg = 5 g

Write whether you multiply or divide to change each measurement. Then complete.

1. 8.74 g = _____ mg

2. 1,900 g = _____ kg

3. 6.21 L = _____ mL

4. _____ m = 5.6 km

5. 226 cm = _____ m

6. _____ g = 5.114 kg

7. 890 mL = _____ L

8. _____ g = 900 mg

9. _____ cm = 6.1 m

10. 0.6 kg = _____ g

11. 0.5 km = _____ m

12. _____ mm = 1.1 m

13. _____ cm = 40 mm

14. 53.7 mL = _____ L

15. 25 cm = _____ mm

4-9 Practice

Integration: Measurement
Changing Metric Units

Write whether you multiply or divide to change each measurement. Then complete.

1. 3.72 L = _____ mL

2. _____ cm = 9.75 m

3. _____ kg = 6.8 g

4. 0.018 kg = _____ g

5. 149 cm = _____ m

6. _____ m = 524 cm

7. _____ g = 0.56 kg

8. 3 mm = _____ cm

9. 2.4 m = _____ mm

10. _____ mg = 6.7 g

11. _____ mL = 9.3 L

12. 0.89 m = _____ cm

13. 0.085 g = _____ mg

14. _____ m = 4,600 mm

15. _____ km = 7,124 m

16. 205 g = _____ kg

17. 609 mg = _____ g

18. _____ mm = 0.0019 m

19. _____ L = 38 mL

20. 720 m = _____ km

21. 150 cm = _____ mm

22. _____ cm = 40 mm

23. _____ m = 7 km

24. _____ mL = 0.0817 L

25. 480 mL = _____ L

26. _____ mm = 53 cm

27. _____ g = 3,020 mg

28. 26 km = _____ m

29. 6.1 mm = _____ m

30. 3.904 L = _____ mL

5-1   Study Guide

Divisibility Patterns

The following rules will help you determine if a number is divisible by
2, 3, 5, 6, 9, or 10.

A number is divisible by:

- 2 if the ones digit is divisible by 2.
- 3 if the sum of the digits is divisible by 3.
- 5 if the ones digit is 0 or 5.
- 6 if the number is divisible by 2 and 3.
- 9 if the sum of the digits is divisible by 9.
- 10 if the ones digit is 0.

Example   **Is 1,120 divisible by 2, 3, 5, 6, 9, or 10?**

2:   Yes. The ones digit 0, is divisible by 2.

3:   No. The sum of the digits, $1 + 1 + 2 + 0$ or 4, is not
divisible by 3.

5:   Yes. The ones digit is 0.

6:   No. The number is divisible by 2, but not by 3.

9:   No. The sum of the digits, $1 + 1 + 2 + 0$ or 4, is not
divisible by 9.

10: Yes. The ones digit is 0. 1,120 is divisible by 10.

1,120 is divisible by 2, 5, and 10.

Determine whether the first number is divisible by the second number.

1. 785; 3   **2.** 655; 5   **3.** 415; 10

4. 819; 9   **5.** 772; 2   **6.** 652; 5

7. 1,180; 10   **8.** 8,764; 9   **9.** 669; 3

10. 4,655; 3   **11.** 6,202; 9   **12.** 3,704; 2

5-1 Practice

Divisibility Patterns

Determine whether the first number is divisible by the second number.

1. 527; 3

2. 1,048; 6

3. 693; 9

4. 1,974; 2

5. 305; 10

6. 860; 5

7. 4,672; 9

8. 2,310; 6

9. 816; 3

10. 13,509; 5

11. 2,847; 2

12. 192; 6

State whether each number is divisible by 2, 3, 5, 6, 9 or 10.

13. 36

14. 450

15. 192

16. 87

17. 264

18. 1,251

19. 890

20. 738

21. 2,601

22. 675

23. 498

24. 2,019

Use mental math skills, paper and pencil, or a calculator to find a number that satisfies the given conditions.

25. a three-digit number divisible by both 2 and 9

26. a number divisible by both 5 and 6

27. a four-digit number divisible by 2, 3, and 9

28. a three-digit number divisible by 5, 6, and 9

29. a four-digit number divisible by 9 and 10

30. a five-digit number divisible by 5 and 9

5-2 Study Guide

Prime Factorization

A **prime number** has exactly two factors, 1 and the number itself.

Example 1 13 is a prime number. It has only 1 and 13 as factors.

A **composite number** has more than two factors.
A composite number may be written as the product of prime numbers.

Example 2 Find the prime factorization of 48.

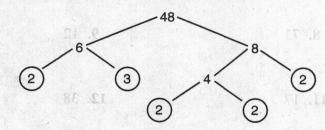

Factor 48.

Factor 6 and 8.

Factor 4.

Circle the prime numbers.

Write the prime numbers in order from least to greatest.
Check to see if the product is 48.

$2 \times 2 \times 2 \times 2 \times 3 = 48$

The numbers 0 and 1 are neither prime nor composite.
0 has an endless number of factors. 1 has only one factor, itself.

Find the prime factorization of each number.

1. 24

2. 25

3. 70

4. 55

5. 44

6. 84

7. 68

8. 121

9. 92

*Mathematics: Applications
and Connections, Course 1*

5-2 Practice

Prime Factorization

Tell whether each number is prime, composite, or neither.

1. 16

2. 0

3. 29

4. 33

5. 47

6. 51

7. 64

8. 73

9. 12

10. 24

11. 17

12. 38

Find the prime factorization of each number.

13. 20

14. 35

15. 54

16. 63

17. 25

18. 88

19. 43

20. 96

21. 64

22. 30

23. 85

24. 57

25. 78

26. 45

27. 84

28. 108

29. 59

30. 69

Study Guide

Greatest Common Factor

The greatest of the common factors of two or more numbers is called the **greatest common factor (GCF)** of the numbers.

Example 1 **Find the GCF of 30 and 42 by making a list.**

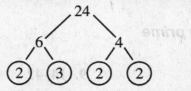

factors of 30: 1, 2, 3, 5, 6, 10, 15, 30
factors of 42: 1, 2, 3,   6, 7, 14, 21, 42

The common factors of 30 and 42 are: 1, 2, 3, and 6.
The greatest common factor of 30 and 42 is 6.

You can also use prime factorization to find the GCF.

Example 2 **Find the GCF of 24 and 40 by using prime factorization.**

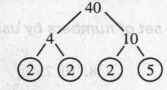

24 and 40 have 2, 2, and 2 as common factors.
The product of the common factors is the GCF.
The GCF of 24 and 40 is 2 × 2 × 2 or 8.

Find the GCF of each set of numbers using either method.

1. 12, 16

2. 18, 24

3. 20, 16

4. 30, 36

5. 35, 49

6. 32, 40

7. 72, 36, 54

8. 63, 42, 21

9. 28, 42, 56

10. 12, 60, 24

11. 35, 28, 21

12. 16, 64, 48

5-3 Practice

Greatest Common Factor

Find the GCF of each set of numbers by making a list.

1. 39, 26

2. 12, 20

3. 18, 27

4. 35, 28

5. 40, 56

6. 36, 48

Find the GCF of each set of numbers by using prime factorization.

7. 24, 30

8. 10, 25

9. 14, 42

10. 54, 72

11. 48, 80

12. 63, 108

Find the GCF of each set of numbers using either method.

13. 6, 21

14. 15, 40

15. 27, 63

16. 56, 14

17. 88, 110

18. 16, 36

19. 24, 64

20. 12, 54

21. 35, 63

22. 60, 84

23. 45, 105

24. 85, 51

25. 16, 24, 56

26. 27, 54, 81

27. 21, 30, 44

Mathematics: Applications and Connections, Course 1

Name _____ **Date** _____

5-4 Study Guide

Simplifying Fractions and Ratios

Fractions that name the same number are called **equivalent fractions.**

Example 1 The same part of each
rectangle is shaded.
The fractions $\frac{3}{4}$ and
$\frac{9}{12}$ are equivalent.

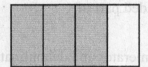

Multiply or divide the numerator and
the denominator of a fraction by the
same number (not zero) to find an
equivalent fraction.

A fraction is in simplest form when
the GCF of the numerator and
denominator is 1.

Examples   2 Solve $\frac{5}{6} = \frac{\square}{18}$.

$$\frac{5}{6} = \frac{\square}{18} \quad \text{Multiply by 3.}$$
(×3 / ×3)

$$\frac{5}{6} = \frac{15}{18}$$

3 Write $\frac{24}{36}$ in simplest form.

$$\frac{24}{36} = \frac{2}{3} \quad \begin{array}{l}\textit{The GCF of 24 and 36}\\ \textit{is 12. Divide by 12.}\end{array}$$
(÷12 / ÷12)

Refer to the figure at the right.

1. What fraction does the shaded part of the figure show?

2. Find the GCF of the numerator and denominator of the fraction.

3. Write the fraction in simplest form.

4. Draw a picture of the fraction in simplified form.

State whether each fraction or ratio is in simplest form. If not, write each fraction or ratio in simplest form.

5. $\frac{9}{21}$ 6. $\frac{12}{13}$ 7. 49:56

8. 48 out of 64   9. $\frac{15}{25}$ 10. $\frac{16}{21}$

11. $\frac{23}{35}$ 12. 32:42   13. $\frac{24}{54}$

© Glencoe/McGraw-Hill

35

*Mathematics: Applications
and Connections, Course 1*

5-4 Practice

Simplifying Fractions and Ratios

Refer to the figure at the right.

1. What fraction does the shaded part of the figure describe?

2. What is the GCF of the numerator and denominator of the fraction?

3. Write the fraction in simplest form.

4. Draw a picture of the fraction in simplest form.

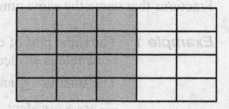

State whether each fraction or ratio is in simplest form. If not, write each fraction or ratio in simplest form.

5. $\frac{9}{14}$ 6. $\frac{6}{21}$ 7. $\frac{45}{72}$ 8. 36:45

9. $\frac{15}{60}$ 10. 70:84   11. $\frac{54}{81}$ 12. 31 out of 61

13. 28: 36   14. $\frac{81}{90}$ 15. 8 to 21   16. $\frac{14}{35}$

17. $\frac{23}{46}$ 18. 45 to 48   19. $\frac{12}{27}$ 20. 17:51

21. $\frac{57}{76}$ 22. $\frac{66}{121}$ 23. 10:25   24. $\frac{19}{29}$

25. 20:28   26. $\frac{49}{56}$ 27. 13 out of 57   28. $\frac{16}{96}$

29. $\frac{63}{108}$ 30. 18:31   31. $\frac{49}{70}$ 32. $\frac{24}{64}$

Mathematics: Applications and Connections, Course 1

5-5 Study Guide

Mixed Numbers and Improper Fractions

The figure shows 2 whole circles plus $\frac{1}{3}$ of a circle. You can write the mixed number $2\frac{1}{3}$ to describe the number of circles.

Mixed numbers may be expressed as **improper fractions**. A fraction in which the numerator is equal to or greater than the denominator is an improper fraction.

Example 1   Express $2\frac{1}{3}$ as an improper fraction.

Multiply the whole number by the denominator.	Add the numerator to the product.	Write the sum over the denominator.
$2 \times 3 = 6$	$6 + 1 = 7$	$\frac{7}{3}$

$$2\frac{1}{3} = \frac{(2 \times 3) + 1}{3} = \frac{7}{3}$$

An improper fraction may be written as a mixed number.

Example 2   Express $\frac{14}{5}$ as a mixed number.

Divide the numerator by the denominator.	Write the quotient as the whole number.	Write the remainder over the denominator as the fraction.
$14 \div 5 = 2 \text{ R } 4$	2	$\frac{4}{5}$

$$\frac{14}{5} = 2\frac{4}{5}$$

Express each mixed number as an improper fraction.

1. $1\frac{2}{5}$ 2. $3\frac{1}{2}$ 3. $7\frac{2}{3}$ 4. $1\frac{7}{8}$

5. $4\frac{3}{4}$ 6. $2\frac{5}{6}$ 7. $5\frac{1}{9}$ 8. $1\frac{2}{7}$

Express each improper fraction as a mixed number.

9. $\frac{12}{7}$ 10. $\frac{17}{9}$ 11. $\frac{17}{5}$ 12. $\frac{25}{6}$

13. $\frac{13}{3}$ 14. $\frac{15}{2}$ 15. $\frac{30}{5}$ 16. $\frac{33}{4}$

Name _____ Date _____

5-5 Practice

Mixed Numbers and Improper Fractions

Express each mixed number as an improper fraction.

1. $1\frac{3}{8}$

2. $2\frac{1}{4}$

3. $3\frac{1}{3}$

4. $1\frac{5}{6}$

5. $6\frac{1}{2}$

6. $3\frac{7}{8}$

7. $2\frac{8}{9}$

8. $10\frac{2}{3}$

9. $5\frac{4}{7}$

10. $4\frac{5}{6}$

11. $9\frac{1}{4}$

12. $8\frac{3}{5}$

Express each improper fraction as a mixed number.

13. $\frac{5}{2}$

14. $\frac{17}{8}$

15. $\frac{9}{5}$

16. $\frac{15}{4}$

17. $\frac{19}{6}$

18. $\frac{27}{4}$

19. $\frac{52}{9}$

20. $\frac{25}{2}$

21. $\frac{37}{5}$

22. $\frac{77}{8}$

23. $\frac{41}{3}$

24. $\frac{31}{7}$

5-6 Study Guide

Integration: Measurement
Length in the Customary System

1 foot (ft) = 12 inches (in.)
1 yard (yd) = 3 feet or 36 inches
1 mile (mi) = 5,280 feet or 1,760 yards

Most inch rulers are divided into eighths, so you can measure to the nearest eighth inch.

Example 1 Draw a line segment measuring $3\frac{3}{8}$ inches.

Use a ruler divided in eighths.

Find $3\frac{3}{8}$ on the ruler.

Draw the line segment from 0 to $3\frac{3}{8}$.

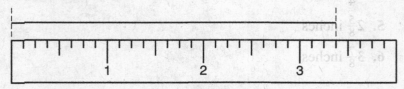

Example 2 Find the length of the toothpick to the nearest eighth inch.

Line up the toothpick with 0 on a ruler divided into eighths.

The toothpick is $2\frac{5}{8}$ inches long.

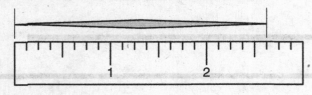

Draw a line segment of each length.

1. $1\frac{1}{2}$ inches

2. $1\frac{1}{8}$ inches

3. $1\frac{1}{4}$ inches

4. $\frac{3}{4}$ inch

5. $1\frac{3}{8}$ inches

6. $1\frac{5}{8}$ inches

Find the length of each object to the nearest half, fourth, or eighth inch.

7.

8.

9.

10.

5-6 Practice

Integration: Measurement
Length in the Customary System

Draw a line segment of each length.

1. $3\frac{1}{2}$ inches

2. $1\frac{3}{8}$ inches

3. $\frac{3}{4}$ inch

4. $4\frac{1}{4}$ inches

5. $2\frac{5}{8}$ inches

6. $3\frac{7}{8}$ inches

Find the length of each line segment to the nearest fourth inch.

7. ────────────

8. ──────────────

9. ───────────────

10. ──────

Find the length of each object to the nearest eighth inch.

11.

12.

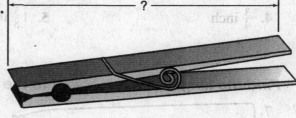

13.

14.

Mathematics: Applications and Connections, Course 1

5-7 Study Guide

Least Common Multiple

A **multiple** of a number is the product of the number and any whole number. The least of the common multiples of two or more numbers, other than zero, is called the **least common multiple (LCM)**.

Examples **1** **Find the LCM of 9 and 12.**

multiples of 9: 0, 9, 18, 27, 36, 45, 54, 63, 72, 81, . . .
multiples of 12: 0, 12, 24, 36, 48, 60, 72, 84, . . .

0, 36, and 72 are common multiples of 9 and 12.
The LCM is 36.

2 **Find the LCM of 4, 6, and 8.**

multiples of 4: 0, 4, 8, 12, 16, 20, 24, 28, 32, 36, 40, 44, 48, . . .
multiples of 6: 0, 6, 12, 18, 24, 30, 36, 42, 48, . . .
multiples of 8: 0, 8, 16, 24, 32, 40, 48, . . .

The LCM of 4, 6, and 8 is 24.

Determine whether the first number is a multiple of the second number.

1. 48, 7   2. 32, 6   3. 144, 4   4. 60, 12

5. 140, 8   6. 90, 18   7. 98, 4   8. 54, 6

Find the LCM for each set of numbers.

9. 12, 20   10. 15, 6   11. 7, 9   12. 15, 9

13. 6, 10   14. 4, 13   15. 18, 27   16. 30, 20

17. 4, 8, 10   18. 3, 5, 7   19. 2, 6, 8   20. 5, 15, 75

5-7

Practice

Study Guide

Least Common Multiple

List the first five multiples of each number.

1. 5   2. 9   3. 11   4. 14

Determine whether the first number is a multiple of the second number.

5. 36; 4   6. 127; 9   7. 42; 3   8. 63; 7

9. 78; 6   10. 144; 8   11. 96; 7   12. 108; 9

Find the LCM for each set of numbers.

13. 3, 7   14. 6, 9   15. 4, 10   16. 5, 6

17. 12, 18   18. 8, 28   19. 6, 14   20. 16, 18

21. 5, 13   22. 12, 15   23. 4, 17   24. 9, 24

25. 8, 13   26. 15, 18   27. 12, 14   28. 7, 13

29. 3, 5, 12   30. 6, 16, 24   31. 12, 18, 24   32. 7, 10, 14

5-8 Study Guide

Comparing and Ordering Fractions

One way to compare fractions is to express them as fractions with the same denominator.

The **least common denominator (LCD)** is the least common multiple of the denominators.

Example   Replace the ◯ with <, >, or = to make a true sentence.

$\frac{5}{8}$ ◯ $\frac{2}{3}$

The LCM of 8 and 3 is 24. Express $\frac{5}{8}$ and $\frac{2}{3}$ as fractions with a denominator of 24.

$$\frac{5}{8} \xrightarrow{\times 3} \frac{15}{24} \qquad \frac{2}{3} \xrightarrow{\times 8} \frac{16}{24}$$

$\frac{15}{24}$ ◯ $\frac{16}{24}$ Compare the numerators. Since $15 < 16$,

$\frac{15}{24} < \frac{16}{24}$. Therefore, $\frac{5}{8} < \frac{2}{3}$.

Find the LCD for each pair of fractions.

1. $\frac{2}{5}, \frac{1}{3}$

2. $\frac{3}{4}, \frac{5}{6}$

3. $\frac{1}{2}, \frac{4}{7}$

Replace each ◯ with <, >, or = to make a true sentence.

4. $\frac{3}{4}$ ◯ $\frac{4}{5}$

5. $\frac{3}{8}$ ◯ $\frac{9}{24}$

6. $\frac{2}{3}$ ◯ $\frac{9}{15}$

7. $\frac{7}{12}$ ◯ $\frac{2}{3}$

8. $\frac{5}{11}$ ◯ $\frac{1}{3}$

9. $\frac{27}{36}$ ◯ $\frac{3}{4}$

Order the following fractions from least to greatest.

10. $\frac{7}{8}, \frac{4}{5}, \frac{3}{4}, \frac{9}{10}$

11. $\frac{1}{3}, \frac{2}{5}, \frac{3}{12}, \frac{3}{10}$

12. $\frac{5}{9}, \frac{1}{2}, \frac{6}{10}, \frac{2}{3}$

13. $\frac{1}{2}, \frac{3}{5}, \frac{2}{7}, \frac{5}{9}$

14. $\frac{1}{10}, \frac{2}{3}, \frac{1}{12}, \frac{5}{6}$

15. $\frac{1}{4}, \frac{7}{10}, \frac{1}{3}, \frac{8}{9}$

5-8 Practice

Comparing and Ordering Fractions

Find the LCD for each pair of fractions.

1. $\frac{4}{5}, \frac{2}{3}$

2. $\frac{5}{8}, \frac{7}{12}$

3. $\frac{1}{2}, \frac{6}{7}$

4. $\frac{1}{6}, \frac{9}{10}$

5. $\frac{3}{4}, \frac{2}{9}$

6. $\frac{5}{12}, \frac{3}{16}$

Replace each $\bigcirc$ with <, >, or = to make a true sentence.

7. $\frac{5}{6} \bigcirc \frac{7}{8}$

8. $\frac{6}{7} \bigcirc \frac{4}{5}$

9. $\frac{3}{9} \bigcirc \frac{1}{3}$

10. $\frac{5}{8} \bigcirc \frac{7}{12}$

11. $\frac{5}{7} \bigcirc \frac{7}{10}$

12. $\frac{2}{3} \bigcirc \frac{3}{4}$

13. $\frac{2}{15} \bigcirc \frac{1}{6}$

14. $\frac{3}{8} \bigcirc \frac{6}{16}$

15. $\frac{5}{12} \bigcirc \frac{2}{5}$

16. $\frac{3}{10} \bigcirc \frac{5}{14}$

17. $\frac{4}{9} \bigcirc \frac{3}{7}$

18. $\frac{1}{6} \bigcirc \frac{2}{12}$

19. $\frac{3}{5} \bigcirc \frac{5}{9}$

20. $\frac{7}{9} \bigcirc \frac{4}{7}$

21. $\frac{9}{10} \bigcirc \frac{11}{12}$

22. $\frac{1}{4} \bigcirc \frac{2}{8}$

23. $\frac{2}{9} \bigcirc \frac{4}{15}$

24. $\frac{8}{9} \bigcirc \frac{7}{8}$

Order the following fractions from least to greatest.

25. $\frac{3}{4}, \frac{2}{5}, \frac{5}{8}, \frac{1}{2}$

26. $\frac{2}{3}, \frac{4}{9}, \frac{5}{6}, \frac{7}{12}$

27. $\frac{1}{3}, \frac{2}{7}, \frac{3}{14}, \frac{1}{6}$

28. $\frac{7}{15}, \frac{3}{5}, \frac{5}{12}, \frac{1}{2}$

29. $\frac{11}{12}, \frac{5}{6}, \frac{3}{4}, \frac{9}{16}$

30. $\frac{4}{5}, \frac{2}{3}, \frac{13}{15}, \frac{7}{9}$

5-9 Study Guide

Writing Decimals as Fractions

A **terminating decimal** can be written as a fraction with a denominator of 10, 100, 1,000, and so on.

Examples **1** **Express 0.5 as a fraction in simplest form.**

$0.5 = \frac{5}{10}$ *Write the decimal as a fraction.*
           Use 10 as the denominator since 0.5 is 5 tenths.

$\frac{5}{10} = \frac{1}{2}$ *Simplify.*

2 **Express 7.005 as a fraction in simplest form.**

$7.005 = 7\frac{5}{1,000}$ *Write the decimal as a mixed number.*
                Use 1,000 as the denominator
                since 0.005 is 5 thousandths.

$7\frac{5}{1,000} = 7\frac{1}{200}$ *Simplify.*

Express each decimal as a fraction or mixed number in simplest form.

1. 0.4   **2.** 0.15   **3.** 0.125   **4.** 0.88

5. 5.008   **6.** 7.8   **7.** 11.25   **8.** 3.525

9. 25.2   **10.** 6.65   **11.** 10.475   **12.** 7.75

13. 12.002   **14.** 5.5   **15.** 8.34   **16.** 2.1

17. Write *fourteen thousandths* as a decimal and as a fraction in simplest form.

18. Write *one and eighty-five hundredths* as a decimal and as a fraction in simplest form.

5-9 Practice

Writing Decimals as Fractions

Express each decimal as a fraction or mixed number in simplest form.

1. 0.2

2. 0.08

3. 0.075

4. 3.56

5. 4.125

6. 10.9

7. 9.35

8. 6.25

9. 8.016

10. 0.055

11. 21.5

12. 7.42

13. 5.006

14. 3.875

15. 1.29

16. 0.004

17. 6.48

18. 2.015

19. 4.95

20. 8.425

21. 9.74

22. 0.824

23. 5.019

24. 1.062

25. 3.96

26. 0.47

27. 20.8

28. 6.45

29. 4.672

30. 0.375

31. Write *twenty-four thousandths* as a decimal and as a fraction in simplest form.

32. Write *twelve and seventy-five hundredths* as a decimal and as a fraction in simplest form.

5-10 Study Guide

Writing Fractions as Decimals

To express a fraction as a decimal, divide the numerator by the denominator. If the division ends with a zero, the decimal is a **terminating decimal**.

Example 1 Express $\frac{7}{8}$ as a decimal.

Use a calculator.

7 ÷ 8 = *0.875*

$\frac{7}{8} = 0.875$

The remainder is 0.
0.875 is a terminating decimal.

Use paper and pencil.

$$\begin{array}{r} 0.875 \\ 8\overline{)7.000} \quad \textit{Annex zeros as needed.} \\ -\,6\,4 \\ \hline 60 \\ -\,56 \\ \hline 40 \\ -\,40 \\ \hline 0 \qquad \frac{7}{8} = 0.875 \end{array}$$

If the decimal repeats rather than terminates, the decimal is a **repeating decimal**. A repeating decimal is written with a bar over the digits that repeat.

Example 2 Express $\frac{5}{6}$ as a decimal.

Use a calculator.

5 ÷ 6 = *0.8333333*

$\frac{5}{6} = 0.8333333 \ldots$

The digit 3 repeats.

$\frac{5}{6} = 0.8\overline{3}$

Use paper and pencil.

$$\begin{array}{r} 0.833 \\ 6\overline{)5.000} \\ -\,4\,8 \\ \hline 20 \\ -\,18 \\ \hline 20 \\ -\,18 \\ \hline 2 \qquad \frac{5}{6} = 0.8\overline{3} \end{array}$$

Express each fraction or mixed number as a decimal. Use bar notation to show a repeating decimal.

1. $\frac{3}{5}$ 2. $\frac{2}{3}$ 3. $\frac{1}{8}$ 4. $\frac{3}{4}$

5. $4\frac{1}{2}$ 6. $3\frac{3}{10}$ 7. $\frac{4}{11}$ 8. $5\frac{5}{9}$

9. $1\frac{3}{8}$ 10. $4\frac{1}{3}$ 11. $\frac{5}{12}$ 12. $3\frac{1}{5}$

13. $\frac{2}{9}$ 14. $6\frac{4}{5}$ 15. $8\frac{1}{4}$ 16. $1\frac{9}{10}$

5-10 Practice

Writing Fractions as Decimals

Write each repeating decimal using bar notation.

1. 0.22222 . . .

2. 0.41666 . . .

3. 0.54545 . . . .

4. 0.6363 . . .

5. 0.2727 . . .

6. 0.428571428 . . .

Express each fraction or mixed number as a decimal. Use bar notation to show a repeating decimal.

7. $\frac{4}{9}$

8. $1\frac{7}{18}$

9. $\frac{5}{7}$

10. $2\frac{3}{16}$

11. $6\frac{1}{12}$

12. $\frac{8}{11}$

13. $9\frac{2}{5}$

14. $7\frac{1}{18}$

15. $3\frac{24}{25}$

16. $4\frac{1}{6}$

17. $\frac{6}{7}$

18. $5\frac{8}{9}$

19. $8\frac{2}{3}$

20. $\frac{5}{16}$

21. $\frac{9}{11}$

22. $10\frac{17}{20}$

23. $2\frac{11}{18}$

24. $6\frac{2}{7}$

25. $14\frac{5}{8}$

26. $\frac{3}{13}$

27. $7\frac{9}{10}$

28. $\frac{7}{12}$

29. $\frac{11}{16}$

30. $1\frac{5}{9}$

6-1 Study Guide

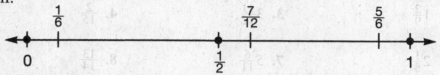

Rounding Fractions and Mixed Numbers

Use the guidelines to round fractions and mixed numbers to the nearest half.

(number line showing $\frac{1}{6}$, $\frac{7}{12}$, $\frac{5}{6}$ above, and 0, $\frac{1}{2}$, 1 below)

- If the numerator is almost as great as the denominator, round the number up to the next whole number. $\frac{5}{6}$ rounds to 1.

- If the numerator is about half of the demominator, round the fraction to $\frac{1}{2}$. $\frac{7}{12}$ rounds to $\frac{1}{2}$.

- If the numerator is much less than the denominator, round the number down to the next whole number. $\frac{1}{6}$ rounds to 0.

Round up when it is better for a measure to be too large than too small. Round down when it is better for a measure to be too small than too large. For example, if you have a 1-gallon bucket and a 2-gallon bucket, choose the 2-gallon bucket to mix $1\frac{1}{3}$ gallons of plant food. If you are allowed 38 pounds of luggage for a trip, pack $36\frac{1}{2}$ pounds rather than $39\frac{1}{2}$ pounds.

Round each number to the nearest half.

1. $\frac{15}{16}$ 2. $4\frac{1}{7}$ 3. $\frac{9}{17}$ 4. $\frac{2}{5}$

5. $\frac{1}{10}$ 6. $9\frac{2}{9}$ 7. $5\frac{4}{7}$ 8. $10\frac{19}{24}$

Tell whether each number should be rounded up, down, or to the nearest half unit.

9. the size of shoe you buy

10. the weight of a package you are mailing

Find the length of each line segment to the nearest half inch.

11. ▬▬▬▬▬▬▬   12. ▬▬▬▬▬▬▬▬▬▬

Practice

Study Guide 6-1

Rounding Fractions and Mixed Numbers

Round each number to the nearest half.

1. $\frac{4}{7}$

2. $1\frac{2}{9}$

3. $3\frac{11}{12}$

4. $\frac{2}{15}$

5. $6\frac{7}{13}$

6. $2\frac{1}{5}$

7. $5\frac{6}{11}$

8. $\frac{13}{14}$

9. $7\frac{3}{19}$

10. $\frac{1}{10}$

11. $9\frac{17}{20}$

12. $4\frac{12}{25}$

13. $8\frac{6}{7}$

14. $1\frac{5}{12}$

15. $\frac{1}{18}$

16. $3\frac{8}{9}$

17. $\frac{9}{16}$

18. $2\frac{11}{13}$

19. $5\frac{3}{20}$

20. $7\frac{9}{11}$

21. $10\frac{1}{7}$

22. $\frac{13}{15}$

23. $6\frac{4}{25}$

24. $8\frac{9}{19}$

Tell whether each number should be rounded up, down, or to the nearest half unit.

25. the length of a strip of wallpaper to hang on a wall $7\frac{5}{6}$ feet high

26. the width of a turntable to fit into a stereo cabinet with a width of $18\frac{2}{5}$ inches

27. the length of a ribbon needed for a bow on a birthday present, accurate to $\frac{1}{2}$ inch

28. the capacity of a punch bowl needed to hold $4\frac{5}{8}$ gallons of punch

Find the length of each line segment to the nearest one-half inch.

29. ▬▬▬▬▬

30. ▬▬▬▬▬▬▬▬▬

31. ▬▬▬▬▬▬▬▬

32. ▬▬▬▬▬▬▬

6-2 Study Guide

Estimating Sums and Differences

Round fractions to the nearest half to estimate sums and differences.

Examples   Estimate.

1 $\frac{9}{16} + \frac{1}{14}$

$\frac{9}{16}$ rounds to $\frac{1}{2}$.

$\frac{1}{14}$ rounds to 0.

Add: $\frac{1}{2} + 0 = \frac{1}{2}$

$\frac{9}{16} + \frac{1}{14}$ is about $\frac{1}{2}$.

2 $\frac{9}{10} - \frac{2}{5}$

$\frac{9}{10}$ rounds to 1.

$\frac{2}{5}$ rounds to $\frac{1}{2}$.

Subtract: $1 - \frac{1}{2} = \frac{1}{2}$

$\frac{9}{10} - \frac{2}{5}$ is about $\frac{1}{2}$.

Round mixed numbers to the nearest whole number to estimate sums and differences.

Examples   Estimate.

3 $8\frac{5}{6} + 6\frac{1}{5}$

$8\frac{5}{6}$ rounds to 9.

$6\frac{1}{5}$ rounds to 6.

Add: $9 + 6 = 15$

$8\frac{5}{6} + 6\frac{1}{5}$ is about 15.

4 $12\frac{5}{8} - 4\frac{1}{2}$

$12\frac{5}{8}$ rounds to 13.

$4\frac{1}{2}$ rounds to 5.

Subtract: $13 - 5 = 8$

$12\frac{5}{8} - 4\frac{1}{2}$ is about 8.

Estimate.

1. $\frac{9}{11} + \frac{3}{8}$

2. $\frac{11}{13} - \frac{7}{9}$

3. $4\frac{2}{3} + 15\frac{4}{5}$

4. $6\frac{1}{3} - 2\frac{5}{8}$

5. $\frac{3}{10} + \frac{3}{5}$

6. $\frac{15}{19} - \frac{3}{7}$

7. $10\frac{3}{4} + 2\frac{1}{3}$

8. $19\frac{1}{2} - 2\frac{5}{8}$

9. $\frac{6}{7} + \frac{1}{8}$

10. $\frac{11}{20} - \frac{1}{4}$

11. $3\frac{3}{16} + 7\frac{7}{10}$

12. $9\frac{1}{4} - 6\frac{2}{15}$

13. $2\frac{12}{13} + \frac{11}{15}$

14. $8\frac{7}{10} - \frac{1}{2}$

15. $\frac{1}{20} + 5\frac{2}{3}$

6-2 Practice

Estimating Sums and Differences

Round each fraction or mixed number to the nearest whole number.

1. $3\frac{7}{9}$

2. $2\frac{2}{5}$

3. $\frac{9}{16}$

4. $5\frac{3}{4}$

5. $\frac{3}{8}$

6. $7\frac{5}{12}$

7. $4\frac{7}{10}$

8. $\frac{6}{7}$

Estimate.

9. $3\frac{4}{9} + 2\frac{3}{5}$

10. $\frac{5}{6} - \frac{1}{7}$

11. $4\frac{7}{12} - \frac{11}{20}$

12. $\frac{9}{10} + \frac{7}{8}$

13. $6\frac{2}{7} - 1\frac{9}{14}$

14. $\frac{11}{12} - \frac{7}{16}$

15. $7\frac{2}{3} + \frac{5}{9}$

16. $\frac{5}{8} + \frac{4}{5}$

17. $9\frac{11}{16} - 3\frac{1}{4}$

18. $8\frac{3}{10} + 5\frac{13}{15}$

19. $6\frac{7}{20} - \frac{1}{5}$

20. $\frac{7}{15} + 2\frac{2}{9}$

21. $\frac{1}{8} + \frac{15}{16}$

22. $12\frac{11}{14} - 7\frac{5}{7}$

23. $\frac{8}{9} - \frac{8}{15}$

24. $8\frac{1}{6} + \frac{14}{15}$

25. $2\frac{3}{7} + 1\frac{5}{14}$

26. $4\frac{1}{3} - \frac{1}{10}$

27. Estimate the sum of $1\frac{5}{9}$ and $4\frac{1}{2}$.

28. Estimate the difference of $\frac{9}{10}$ and $\frac{11}{20}$.

Mathematics: Applications and Connections, Course 1

6-4 Study Guide

Adding and Subtracting Fractions with Unlike Denominators

To find the sum or difference of two fractions with unlike denominators, write equivalent fractions with a common denominator. Then add or subtract.

Examples

1 Find $\frac{3}{4} + \frac{5}{6}$.

$\frac{3}{4} + \frac{5}{6} = \frac{9}{12} + \frac{10}{12}$ *The LCM of 4 and 6 is 12.*
Rename the fractions with 12 as
     $= \frac{9 + 10}{12}$ *the denominator.*

     $= \frac{19}{12}$ or $1\frac{7}{12}$

2 Find $\frac{2}{3} - \frac{3}{5}$.

$\frac{2}{3} - \frac{3}{5} = \frac{10}{15} - \frac{9}{15}$ *The LCM of 3 and 6 is 15.*
Rename the fractions with 15
     $= \frac{10 - 9}{15}$ *as the denominator.*

     $= \frac{1}{15}$

Find the LCD for each pair of fractions.

1. $\frac{1}{6}, \frac{2}{3}$ **2.** $\frac{1}{2}, \frac{2}{5}$ **3.** $\frac{7}{8}, \frac{5}{6}$ **4.** $\frac{4}{9}, \frac{1}{3}$

5. $\frac{5}{8}, \frac{1}{3}$ **6.** $\frac{3}{10}, \frac{4}{15}$ **7.** $\frac{5}{12}, \frac{1}{2}$ **8.** $\frac{13}{20}, \frac{2}{5}$

Add or subtract. Write each answer in simplest form.

9. $\frac{1}{6} + \frac{1}{2}$ **10.** $\frac{2}{3} - \frac{1}{2}$ **11.** $\frac{1}{4} + \frac{7}{8}$ **12.** $\frac{9}{10} - \frac{3}{5}$

13. $\frac{4}{5} + \frac{1}{12}$ **14.** $\frac{11}{15} - \frac{1}{3}$ **15.** $\frac{1}{9} + \frac{1}{6}$ **16.** $\frac{1}{2} - \frac{7}{16}$

17. $\frac{3}{10} + \frac{4}{5}$ **18.** $\frac{4}{5} - \frac{1}{6}$ **19.** $\frac{2}{3} + \frac{1}{2}$ **20.** $\frac{7}{8} - \frac{4}{9}$

6-4 Practice

Adding and Subtracting Fractions with Unlike Denominators

Find the LCD for each pair of fractions.

1. $\frac{3}{8}, \frac{5}{12}$

2. $\frac{2}{9}, \frac{4}{15}$

3. $\frac{1}{6}, \frac{9}{14}$

4. $\frac{2}{3}, \frac{3}{16}$

5. $\frac{4}{5}, \frac{5}{6}$

6. $\frac{7}{10}, \frac{5}{8}$

Add or subtract. Write the answer in simplest form.

7. $\frac{3}{4} + \frac{5}{6}$

8. $\frac{7}{8} - \frac{2}{3}$

9. $\frac{4}{7} - \frac{1}{2}$

10. $\frac{8}{9} - \frac{5}{12}$

11. $\frac{2}{3} + \frac{3}{5}$

12. $\frac{6}{7} + \frac{1}{4}$

13. $\frac{1}{6} + \frac{13}{15}$

14. $\frac{9}{16} - \frac{5}{12}$

15. $\frac{1}{3} + \frac{5}{7}$

16. $\frac{7}{12} - \frac{9}{20}$

17. $\frac{11}{12} - \frac{3}{4}$

18. $\frac{5}{6} + \frac{9}{10}$

19. $\frac{3}{8} - \frac{2}{7}$

20. $\frac{8}{9} - \frac{2}{3}$

21. $\frac{4}{5} + \frac{2}{9}$

22. $\frac{1}{2} - \frac{3}{11}$

23. $\frac{7}{9} - \frac{3}{4}$

24. $\frac{5}{8} + \frac{9}{16}$

25. $\frac{1}{3} + \frac{6}{13}$

26. $\frac{5}{12} - \frac{2}{5}$

27. $\frac{3}{5} + \frac{3}{8}$

28. What is the sum of the fractions $\frac{1}{4}$, $\frac{3}{8}$, and $\frac{5}{14}$?

29. What is the difference of the fractions $\frac{7}{12}$ and $\frac{8}{15}$?

6-5 Study Guide

Adding and Subtracting Mixed Numbers

Follow these steps to add and subtract mixed numbers.

Examples

	Write equivalent fractions with a common denominator. Add or subtract the fractions.	Add or subtract the whole numbers. Rename and simplify if necessary.

1 $4\frac{7}{8}$ $+ 3\frac{1}{2}$ $\longrightarrow$ $4\frac{7}{8}$ $+ 3\frac{4}{8}$ $\overline{\frac{11}{8}}$ $\longrightarrow$ $4\frac{7}{8}$ $+ 3\frac{4}{8}$ $\overline{7\frac{11}{8}}$ $7 + 1\frac{3}{8} = 8\frac{3}{8}$

2 $5\frac{5}{6}$ $- 3\frac{1}{2}$ $\longrightarrow$ $5\frac{5}{6}$ $- 3\frac{3}{6}$ $\overline{\frac{2}{6}}$ $\longrightarrow$ $5\frac{5}{6}$ $- 3\frac{3}{6}$ $\overline{2\frac{2}{6} = 2\frac{1}{3}}$

Add or subtract. Write each answer in simplest form.

1. $6\frac{1}{4} + 2\frac{1}{4}$

2. $7\frac{7}{9} - 4\frac{2}{9}$

3. $8\frac{2}{3} + 2\frac{1}{3}$

4. $6\frac{5}{7} - 5\frac{2}{7}$

5. $10\frac{1}{2} + 4\frac{1}{8}$

6. $12\frac{5}{6} - 3\frac{1}{3}$

7. $7\frac{1}{10} + 2\frac{1}{5}$

8. $9\frac{1}{2} - 5\frac{1}{6}$

9. $5\frac{3}{4} + 2\frac{5}{8}$

10. $18\frac{3}{4} - 6\frac{3}{4}$

11. $5\frac{6}{7} + 4\frac{2}{3}$

12. $9\frac{3}{4} - 2\frac{1}{6}$

6-5 Practice

Adding and Subtracting Mixed Numbers

Add or subtract. Write the answer in simplest form.

1. $2\frac{3}{7} + 4\frac{2}{7}$

2. $6\frac{2}{3} + 3\frac{4}{9}$

3. $8\frac{7}{12} - 5\frac{5}{12}$

4. $10\frac{3}{5} - 2\frac{1}{2}$

5. $6\frac{5}{6} + \frac{3}{8}$

6. $9\frac{4}{5} + 2\frac{2}{3}$

7. $7\frac{15}{16} - 3\frac{7}{16}$

8. $5\frac{8}{9} - 3\frac{1}{6}$

9. $8\frac{3}{4} + 6\frac{2}{5}$

10. $13\frac{3}{10} - 8\frac{2}{15}$

11. $11\frac{2}{3} - 3\frac{4}{7}$

12. $3\frac{5}{9} + 7\frac{4}{9}$

13. $9\frac{1}{2} + 3\frac{8}{9}$

14. $14\frac{7}{8} - 8\frac{3}{4}$

15. $12\frac{5}{12} - 5\frac{7}{18}$

16. $8\frac{7}{10} + 7\frac{9}{10}$

17. $13\frac{7}{12} - 9\frac{1}{4}$

18. $2\frac{5}{7} + 7\frac{1}{2}$

19. $14\frac{2}{3} + \frac{5}{6}$

20. $7\frac{11}{12} - 6\frac{5}{8}$

21. $10\frac{5}{6} - 3\frac{5}{6}$

22. $8\frac{2}{7} + 3\frac{4}{5}$

23. $5\frac{8}{11} + 5\frac{1}{3}$

24. $9\frac{15}{16} - 7\frac{3}{8}$

25. $4\frac{9}{10} - 3\frac{2}{5}$

26. $8\frac{2}{3} + 9\frac{5}{8}$

27. $7\frac{19}{20} - 4\frac{7}{10}$

Solve each equation. Write the solution in simplest form.

28. $x = 2\frac{5}{6} + 3\frac{2}{7}$

29. $16\frac{5}{12} - 7\frac{2}{9} = n$

30. $a = 18\frac{3}{20} - 5\frac{1}{15}$

31. $12\frac{7}{15} - 5\frac{1}{3} = c$

6-6 Study Guide

Subtracting Mixed Numbers with Renaming

Sometimes it is necessary to rename a mixed number as an improper fraction before you can subtract.

Examples **1**

$$6\frac{1}{2} \qquad 6\frac{2}{4}$$

You cannot subtract $\frac{3}{4}$ from $\frac{2}{4}$.

Rename $6\frac{2}{4}$ as $5\frac{6}{4}$.

$$-2\frac{3}{4} \longrightarrow -2\frac{3}{4} \qquad\qquad \longrightarrow -2\frac{3}{4}$$
$$3\frac{3}{4}$$

Then subtract.

2

$$8 \longrightarrow 7\frac{8}{8}$$

Rename 8 as $7\frac{8}{8}$.

$$-4\frac{5}{8} \longrightarrow -4\frac{5}{8}$$
$$3\frac{3}{8}$$

Then subtract.

Complete.

1. $7\frac{5}{6} = \boxed{}\frac{11}{6}$

2. $4\frac{3}{4} = 3\frac{\square}{4}$

3. $2\frac{3}{8} = 1\frac{\square}{8}$

4. $9\frac{3}{5} = \boxed{}\frac{8}{5}$

5. $10\frac{1}{3} = 9\frac{\square}{3}$

6. $15 = 14\frac{\square}{2}$

7. $20\frac{5}{12} = 19\frac{\square}{12}$

8. $13 = 12\frac{\square}{7}$

9. $6\frac{2}{5} = \boxed{}\frac{7}{5}$

Subtract. Write each answer in simplest form.

10. $5\frac{1}{3} - 3\frac{2}{3}$

11. $12\frac{1}{6} - 7\frac{5}{6}$

12. $8\frac{3}{8} - 3\frac{5}{8}$

13. $9\frac{1}{2} - 4\frac{3}{4}$

14. $12 - 1\frac{2}{5}$

15. $8\frac{1}{2} - \frac{7}{8}$

16. $15\frac{1}{3} - 9\frac{5}{6}$

17. $7\frac{1}{2} - 3\frac{11}{12}$

18. $22 - 10\frac{8}{9}$

Mathematics: Applications and Connections, Course 1

6-6 Practice

Subtracting Mixed Numbers with Renaming

Complete.

1. $6\frac{4}{7} = 5\frac{\square}{7}$ 2. $9\frac{3}{10} = \boxed{}\frac{13}{10}$ 3. $8\frac{5}{12} = \boxed{}\frac{17}{12}$ 4. $3\frac{7}{9} = 2\frac{\square}{9}$

5. $10\frac{9}{14} = 9\frac{\square}{14}$ 6. $7\frac{3}{11} = 6\frac{\square}{11}$ 7. $12\frac{7}{8} = \boxed{}\frac{15}{8}$ 8. $14\frac{13}{15} = 13\frac{\square}{15}$

Subtract. Write the answer in simplest form.

9. $\begin{array}{r} 4\frac{5}{7} \\ -\ 1\frac{6}{7} \\ \hline \end{array}$ 10. $\begin{array}{r} 7\frac{1}{6} \\ -\ 3\frac{5}{6} \\ \hline \end{array}$ 11. $\begin{array}{r} 10 \\ -\ 5\frac{1}{4} \\ \hline \end{array}$ 12. $\begin{array}{r} 12\frac{5}{8} \\ -\ 3\frac{3}{4} \\ \hline \end{array}$

13. $\begin{array}{r} 15\frac{1}{6} \\ -\ 9\frac{4}{9} \\ \hline \end{array}$ 14. $\begin{array}{r} 17\frac{1}{5} \\ -\ 7\frac{2}{3} \\ \hline \end{array}$ 15. $\begin{array}{r} 15\frac{5}{14} \\ -\ 8\frac{4}{7} \\ \hline \end{array}$ 16. $\begin{array}{r} 13\frac{1}{2} \\ -\ 7\frac{4}{5} \\ \hline \end{array}$

17. $\begin{array}{r} 10\frac{1}{6} \\ -\ 2\frac{3}{8} \\ \hline \end{array}$ 18. $\begin{array}{r} 2\frac{1}{7} \\ -\ \frac{2}{3} \\ \hline \end{array}$ 19. $\begin{array}{r} 14\frac{1}{12} \\ -\ 3\frac{8}{9} \\ \hline \end{array}$ 20. $\begin{array}{r} 15\frac{9}{16} \\ -\ 6\frac{5}{8} \\ \hline \end{array}$

21. $\begin{array}{r} 3\frac{2}{5} \\ -\ 2\frac{3}{4} \\ \hline \end{array}$ 22. $\begin{array}{r} 16\frac{2}{3} \\ -\ 12\frac{11}{12} \\ \hline \end{array}$ 23. $\begin{array}{r} 18\frac{1}{9} \\ -\ 8\frac{5}{18} \\ \hline \end{array}$ 24. $\begin{array}{r} 9\frac{7}{15} \\ -\ 6\frac{4}{5} \\ \hline \end{array}$

25. $\begin{array}{r} 12 \\ -\ 5\frac{7}{11} \\ \hline \end{array}$ 26. $\begin{array}{r} 14\frac{2}{7} \\ -\ 9\frac{3}{4} \\ \hline \end{array}$ 27. $\begin{array}{r} 16\frac{3}{4} \\ -\ 5\frac{5}{6} \\ \hline \end{array}$ 28. $\begin{array}{r} 8\frac{1}{20} \\ -\ 7\frac{3}{10} \\ \hline \end{array}$

29. If you subtract $4\frac{2}{5}$ from $8\frac{3}{8}$, what is the result?

30. Find the difference of $5\frac{1}{6}$ and $2\frac{5}{12}$.

6-7 Study Guide

Integration: Measurement
Adding and Subtracting Measures of Time

To add measures of time, add the seconds, add the minutes, and add the hours. Rename if necessary.

Remember: 1 hour (h) = 60 minutes (min)
            1 minute (min) = 60 seconds (s)

Example 1   4 h 25 min 40 s
           + 5 h 30 min 25 s
             9 h 55 min 65 s   *Rename 65 s as 1 min 5 s.*

             9 h 55 min + 1 min 5 s = 9 h 56 min 5 s

To subtract measures of time, rename if necessary.
Then subtract seconds, subtract minutes, and subtract hours.

Example 2   7 h 15 min 40 s   *You cannot subtract 20 min from 15 min.*
           − 3 h 20 min 10 s
                   ↓
             6 h 75 min 40 s   Rename 7 h 15 min as 6 h 75 min.
           − 3 h 20 min 10 s   Then subtract.
             3h 55 min 30 s

Complete.

1. 14 min 85 s = _____ min 25 s   **2.** 9 h 5 min = 8 h _____ min

3. 3 h 20 min 7 s = 3 h _____ min _____ s   **4.** 7 h 9 min 25 s = _____ h _____ min 25 s

Add or subtract. Rename if necessary.

5.   6 h 20 min   **6.**   35 min 45 s   **7.**   12 h 15 s
     − 3 h 17 min   **+ 12 min 12 s**   **+ 10 h 55 s**

8.   9 h 45 min 10 s   **9.**   1 h 55 min 12 s   **10.**   7 h 20 min
     − 3 h 30 min 50 s   **+ 3 h 25 min 34 s**   **− 2 h 9 min 10 s**

6-7 Practice

Integration: Measurement
Adding and Subtracting Measures of Time

Complete.

1. 9 h 24 min = 8 h _____ min

2. 15 min 95 s = _____ min 35 s

3. 12 h 108 min = _____ h 48 min

4. 20 min 54 s = 19 min _____ s

5. 3 h 12 min 6 s = _____ h 72 min 6 s

6. 7 h 46 min 15 s = 6 h _____ min 75 s

Add or subtract. Rename if necessary.

7.   9 h 42 min
   − 3 h 18 min

8.   23 min 16 s
   + 12 min 34 s

9.   6 h 38 min
   + 5 h 22 min

10.   18 h 27 min
   − 9 h 52 min

11.   45 min 17 s
   − 39 min 50 s

12.   8 h 49 min
   + 7 h 35 min

13.   21 min 54 s
   + 26 min 19 s

14.   14 h
   − 9 h 43 min

15.   51 min 6 s
   − 37 min 48 s

16.   32 min
   − 14 min 16 s

17.   13 h 41 min
   + 6 h 19 min

18.   38 min 48 s
   + 29 min 12 s

19.   11 h 23 min 6 s
   − 5 h 36 min 29 s

20.   6 h 10 min 47 s
   + 2 h 51 min 28 s

21.   20 h
   − 8 h 33 min 18 s

Find the elapsed time.

22. 6:35 A.M. to 9:55 A.M.

23. 12:20 P.M. to 3:05 P.M.

24. 10:45 A.M. to 5:25 P.M.

25. 8:10 P.M. to 1:15 A.M.

26. 11:50 A.M. to 7:25 P.M.

27. 2:40 P.M. to 4:35 P.M.

7-1

Study Guide

Estimating Products

One way to estimate products is using **compatible numbers.**
Compatible numbers are easy to divide mentally.

Example 1 Estimate $\frac{3}{5} \times 9$.

$\frac{1}{5} \times 10 = 2$ *For 9, the nearest multiple of 5 is 10. $\frac{1}{5}$ of 10 is 2.*

$\frac{3}{5} \times 10 = 6$ *Since $\frac{1}{5} \times 10 = 2$, it follows that $\frac{3}{5}$ of 10 is 3×2 or 6.*

So, $\frac{3}{5} \times 9$ is about 6.

You can also estimate products by rounding. Round fractions to 0, $\frac{1}{2}$ or 1.
Round mixed numbers to the nearest whole number.

Examples   **2**   Estimate $\frac{5}{6} \times \frac{3}{7}$.

$\frac{5}{6} \times \frac{3}{7}$ *Think of a number line.*
              $\frac{5}{6}$ rounds to 1.

$1 \times \frac{3}{7} = \frac{3}{7}$ *So, $\frac{5}{6} \times \frac{3}{7}$ is about $\frac{3}{7}$.*

   3   Estimate $9\frac{9}{11} \times 5\frac{1}{5}$.

$9\frac{9}{11} \times 5\frac{1}{5}$.   *Round $9\frac{9}{11}$ to 10.*

                Round $5\frac{1}{5}$ to 5.

$10 \times 5 = 50$ *So, $9\frac{9}{11} \times 5\frac{1}{5}$ is about 50.*

Round each fraction to 0, $\frac{1}{2}$ or 1.

1. $\frac{6}{13}$ **2.** $\frac{3}{8}$ **3.** $\frac{9}{10}$ **4.** $\frac{1}{9}$

Estimate each product.

5. $\frac{1}{5} \times 24$ **6.** $7\frac{2}{7} \times 5\frac{3}{4}$ **7.** $\frac{2}{3} \times 19$ **8.** $\frac{9}{10} \times \frac{3}{5}$

9. $15 \times \frac{1}{4}$ **10.** $8\frac{7}{8} \times 2\frac{9}{10}$ **11.** $\frac{1}{9} \times \frac{1}{12}$ **12.** $\frac{3}{7} \times 27$

7-1 Practice

Estimating Products

Round each fraction to 0, $\frac{1}{2}$, or 1.

1. $\frac{4}{5}$

2. $\frac{1}{6}$

3. $\frac{7}{15}$

4. $\frac{2}{9}$

5. $\frac{6}{7}$

6. $\frac{5}{11}$

Round each mixed number to the nearest whole number.

7. $5\frac{3}{4}$

8. $9\frac{1}{3}$

9. $3\frac{7}{10}$

10. $7\frac{5}{8}$

11. $10\frac{4}{15}$

12. $2\frac{5}{12}$

Estimate each product.

13. $\frac{2}{5} \times 11$

14. $\frac{6}{7} \times \frac{1}{8}$

15. $6\frac{3}{10} \times 4\frac{7}{9}$

16. $\frac{1}{3} \times \frac{8}{15}$

17. $2\frac{1}{2} \times 5\frac{1}{4}$

18. $\frac{4}{7} \times 29$

19. $8\frac{11}{15} \times 3\frac{9}{14}$

20. $\frac{5}{6} \times \frac{2}{7}$

21. $\frac{2}{9} \times 52$

22. $\frac{1}{5} \times \frac{6}{13}$

23. $7\frac{2}{3} \times 9\frac{3}{8}$

24. $\frac{3}{4} \times 35$

25. $12\frac{4}{15} \times 3\frac{9}{10}$

26. $\frac{7}{8} \times 30$

27. $\frac{11}{20} \times \frac{1}{7}$

28. $6\frac{7}{12} \times 8\frac{5}{14}$

29. $\frac{2}{3} \times 35$

30. $\frac{1}{2} \times \frac{5}{9}$

31. Estimate the product of $\frac{3}{16}$ and $\frac{8}{9}$.

32. Estimate the product of $2\frac{4}{11}$ and $16\frac{1}{5}$.

Mathematics: Applications and Connections, Course 1

7-2 Study Guide

Multiplying Fractions

To multiply fractions: Multiply the numerators.   Then multiply the denominators. Write the product in simplest form.

$$\frac{5}{7} \times \frac{3}{5}$$

$$\frac{5}{7} \times \frac{3}{5} = \frac{15}{}$$

$$\frac{5}{7} \times \frac{3}{5} = \frac{15}{35} = \frac{3}{7}$$

To multiply whole numbers and fractions: Rename the whole number as an improper fraction. Multiply the numerators.   Multiply the denominators. Write the product in simplest form.

$$\frac{3}{8} \times 7$$

$$\frac{3}{8} \times \frac{7}{1} = \frac{21}{}$$

$$\frac{3}{8} \times \frac{7}{1} = \frac{21}{8} = 2\frac{5}{8}$$

If the numerator of one fraction and the denominator of the other fraction have a common factor, you can simplify before you multiply.

Example   Find $\frac{8}{11} \times \frac{3}{4}$.   *The GCF of 8 and 4 is 4.*

$$\frac{\overset{2}{8}}{11} \times \frac{3}{\underset{1}{4}} = \frac{6}{11}$$ *Divide the numerator and denominator by 4. Then multiply.*

Find each product. Write in simplest form.

1. $\frac{1}{3} \times \frac{1}{5}$ 2. $\frac{5}{8} \times \frac{1}{2}$ 3. $\frac{4}{9} \times \frac{3}{4}$

4. $6 \times \frac{2}{3}$ 5. $\frac{3}{5} \times 10$ 6. $\frac{2}{3} \times \frac{3}{8}$

7. $\frac{1}{7} \times \frac{1}{7}$ 8. $\frac{2}{9} \times \frac{1}{2}$ 9. $12 \times \frac{5}{6}$

Solve each equation. Write the solution in simplest form.

10. $m = 8 \times \frac{1}{4}$ 11. $\frac{3}{5} \times \frac{5}{6} = n$ 12. $c = \frac{2}{7} \times \frac{1}{3}$

13. $\frac{5}{8} \times 24 = a$ 14. $k = \frac{5}{12} \times \frac{1}{5}$ 15. $\frac{1}{2} \times \frac{1}{5} = t$

16. $e = \frac{6}{7} \times \frac{8}{15}$ 17. $\frac{5}{12} \times 10 = t$ 18. $h = \frac{8}{9} \times \frac{9}{10}$

7-2 Practice

Multiplying Fractions

Find each product. Write in simplest form.

1. $\frac{3}{4} \times \frac{1}{2}$

2. $\frac{1}{3} \times \frac{5}{6}$

3. $\frac{2}{5} \times \frac{3}{7}$

4. $\frac{3}{8} \times 10$

5. $\frac{1}{6} \times \frac{3}{5}$

6. $\frac{1}{4} \times \frac{2}{7}$

7. $\frac{2}{3} \times \frac{5}{8}$

8. $\frac{9}{10} \times \frac{4}{5}$

9. $\frac{7}{8} \times \frac{2}{9}$

10. $16 \times \frac{5}{12}$

11. $\frac{4}{9} \times \frac{1}{8}$

12. $\frac{5}{6} \times \frac{7}{10}$

13. $\frac{1}{5} \times \frac{15}{16}$

14. $\frac{1}{10} \times \frac{4}{7}$

15. $\frac{5}{9} \times 18$

Solve each equation. Write the solution in simplest form.

16. $a = \frac{5}{9} \times \frac{4}{5}$

17. $\frac{5}{7} \times \frac{14}{15} = s$

18. $\frac{7}{18} \times \frac{3}{14} = p$

19. $r = \frac{2}{3} \times \frac{9}{10}$

20. $28 \times \frac{5}{8} = x$

21. $\frac{5}{6} \times \frac{4}{7} = m$

22. $d = \frac{4}{9} \times \frac{15}{16}$

23. $\frac{3}{10} \times \frac{5}{8} = k$

24. $h = \frac{3}{7} \times 35$

25. $n = \frac{3}{20} \times \frac{5}{6}$

26. $\frac{9}{14} \times \frac{7}{12} = z$

27. $f = \frac{5}{12} \times \frac{4}{15}$

28. $\frac{3}{14} \times \frac{2}{9} = c$

29. $\frac{3}{4} \times \frac{8}{9} = t$

30. $y = 15 \times \frac{3}{10}$

31. Find the product of $\frac{4}{5}$ and 30.

32. Evaluate xy if $x = \frac{2}{3}$ and $y = \frac{3}{4}$.

7-3 Study Guide

Multiplying Mixed Numbers

To multiply mixed numbers, express each mixed number as an improper fraction. Then multiply the fractions.

Example Find $7\frac{1}{2} \times 3\frac{1}{3}$.

Estimate: $8 \times 3 = 24$

$7\frac{1}{2} \times 3\frac{1}{3} = \frac{15}{2} \times \frac{10}{3}$ Express the mixed numbers as improper fractions.

$= \frac{\overset{5}{15} \cdot \overset{5}{10}}{\underset{1}{2} \cdot \underset{1}{3}}$ Divide 15 and 3 by the GCF, 3.
Divide 10 and 2 by the GCF, 2.

$= \frac{25}{1}$ or 25

Express each mixed number as an improper fraction.

1. $5\frac{3}{4}$

2. $3\frac{7}{9}$

3. $2\frac{4}{5}$

4. $1\frac{15}{16}$

Find each product. Write in simplest form.

5. $\frac{2}{3} \times 3\frac{1}{2}$

6. $5\frac{3}{4} \times \frac{2}{3}$

7. $9 \times 1\frac{5}{6}$

8. $2\frac{4}{9} \times \frac{4}{11}$

9. $1\frac{1}{4} \times \frac{3}{5}$

10. $2\frac{1}{2} \times 1\frac{1}{5}$

11. $\frac{1}{8} \times 1\frac{1}{3}$

12. $8 \times 1\frac{1}{4}$

Solve each equation. Write the solution in simplest form.

13. $2\frac{1}{2} \times 4 = n$

14. $k = 4\frac{2}{3} \times 1\frac{1}{2}$

15. $\frac{4}{5} \times 1\frac{1}{4} = p$

16. $y = 6 \times 3\frac{1}{3}$

17. $4\frac{1}{2} \times \frac{8}{9} = a$

18. $8\frac{1}{3} \times \frac{3}{5} = r$

7-3 Practice

Multiplying Mixed Numbers

Express each mixed number as an improper fraction.

1. $2\frac{4}{5}$

2. $6\frac{3}{4}$

3. $8\frac{6}{7}$

4. $5\frac{2}{9}$

5. $9\frac{5}{6}$

6. $3\frac{1}{12}$

Find each product. Write in simplest form.

7. $3\frac{1}{5} \times \frac{3}{4}$

8. $9 \times 4\frac{2}{3}$

9. $2\frac{5}{6} \times 4\frac{1}{2}$

10. $\frac{4}{7} \times 3\frac{1}{9}$

11. $1\frac{3}{8} \times 2\frac{2}{7}$

12. $4\frac{1}{6} \times \frac{9}{10}$

13. $3\frac{1}{3} \times 2\frac{1}{4}$

14. $\frac{8}{9} \times 5\frac{1}{7}$

15. $2\frac{5}{8} \times 6$

16. $3\frac{3}{4} \times 2\frac{4}{5}$

17. $\frac{5}{7} \times 4\frac{3}{8}$

18. $20 \times 1\frac{2}{5}$

Solve each equation. Write the solution in simplest form.

19. $2\frac{4}{9} \times \frac{6}{11} = s$

20. $p = 1\frac{1}{8} \times 3\frac{3}{7}$

21. $\frac{6}{7} \times 2\frac{5}{12} = x$

22. $d = 14 \times 1\frac{3}{4}$

23. $5\frac{2}{5} \times \frac{8}{9} = t$

24. $3\frac{3}{5} \times 2\frac{2}{9} = a$

25. $r = 1\frac{4}{5} \times 3\frac{4}{7}$

26. $n = \frac{2}{3} \times 5\frac{1}{6}$

27. $1\frac{5}{14} \times \frac{7}{8} = y$

28. $k = 2\frac{3}{8} \times 16$

29. $5\frac{1}{4} \times 2\frac{1}{3} = b$

30. $m = \frac{7}{9} \times 5\frac{5}{8}$

7-4 Study Guide

Integration: Geometry
Circles and Circumferences

A **circle** is all of the points in a plane that are the same distance
from a point called the **center**. The **diameter** (d) is the distance
across the circle through its center. The **radius** (r) is the distance
from the center to any point on the circle. The **circumference** (C)
is the distance around the circle.

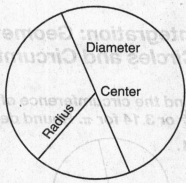

Examples

1 **Find the circumference of a circle with a diameter of 7.5 inches.**

$C = \pi d$ *Formula for circumference*

$\approx 3.14 \cdot 7.5$ *Replace π with 3.14 and d with 7.5.*

3.14 ☒ 7.5 ▭ *23.55*

The circumference of the circle is about 23.55 inches.

2 **Find the circumference of a circle with a radius of 21 meters.**

$C = 2\pi r$ *Formula for circumference*

$= 2 \cdot \frac{22}{7} \cdot 21$ *Replace π with $\frac{22}{7}$ and r with 21.*

$= \frac{2}{1} \cdot \frac{22}{7} \cdot \frac{21^3}{1}$ *Divide 7 and 21 by the GCF, 7.*

$= \frac{132}{1}$ or 132   *Simplify.*

The circumference of the circle is about 132 meters.

**Find the circumference of each circle to the nearest tenth.
Use $\frac{22}{7}$ or 3.14 for π.**

1.
25 ft

2.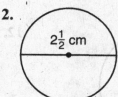
$2\frac{1}{2}$ cm

3.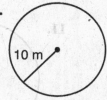
10 m

4.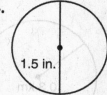
1.5 in.

5. $d = 11$ km

6. $r = 24$ mi

7. $d = 12$ m

8. $r = 30\frac{3}{4}$ ft

9. $r = 4\frac{2}{3}$ yd

10. $d = 15$ km

11. $r = 50$ ft

12. $d = 4.1$ m

7-4 Practice

Practice

Integration: Geometry
Circles and Circumferences

Find the circumference of each circle shown or described. Use $\frac{22}{7}$ or 3.14 for π. Round decimal answers to the nearest tenth.

1.
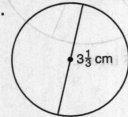
$3\frac{1}{3}$ cm

2.
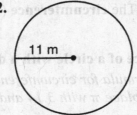
11 m

3.
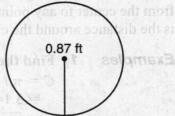
0.87 ft

4.
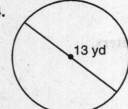
13 yd

5.
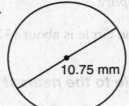
$3\frac{3}{4}$ mi

6.
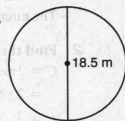
18.5 m

7.
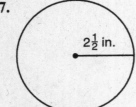
$2\frac{1}{2}$ in.

8.
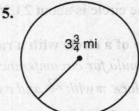
10.75 mm

9.
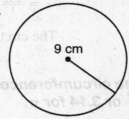
9 cm

10.
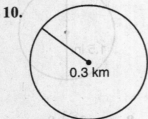
0.3 km

11.
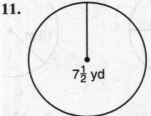
$7\frac{1}{2}$ yd

12.
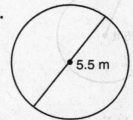
5.5 m

13. $d = 4.1$ ft

14. $r = 4\frac{2}{3}$ mi

15. $r = 0.7$ km

16. $d = 16$ cm

17. $r = 22$ mm

18. $d = 1\frac{1}{4}$ in.

© Glencoe/McGraw-Hill

52

Mathematics: Applications and Connections, Course 1

7-5 Study Guide

Dividing Fractions

Two numbers are **reciprocals** if their product is 1.

$\frac{1}{3}$ and 3 are reciprocals. $\frac{17}{18}$ and $\frac{18}{17}$ are reciprocals.

$\frac{1}{3} \times 3 = 1$ $\frac{17}{18} \times \frac{18}{17} = 1$

You use reciprocals to divide fractions.
To divide by a fraction, multiply by its reciprocal.

Example   Find $\frac{5}{6} \div \frac{2}{3}$.

$\frac{5}{6} \div \frac{2}{3} = \frac{5}{6} \times \frac{3}{2}$ *Multiply by the reciprocal of $\frac{2}{3}$.*

$= \frac{5}{6}^{\,2} \times \frac{3}{2}^{\,1}$ *Divide 3 and 6 by the GCF, 3.*

$= \frac{5}{2} \times \frac{1}{2} = \frac{5}{4}$ *Multiply the numerators.*
     Multiply the denominators.

$= 1\frac{1}{4}$

Find the reciprocal of each number.

1. $\frac{3}{4}$
2. $\frac{5}{8}$
3. 9
4. $\frac{12}{13}$

Find each quotient. Write in simplest form.

5. $\frac{1}{2} \div \frac{3}{4}$
6. $\frac{4}{5} \div \frac{1}{10}$
7. $\frac{3}{8} \div \frac{3}{4}$
8. $\frac{7}{9} \div \frac{1}{3}$

9. $\frac{14}{15} \div 7$
10. $\frac{5}{12} \div \frac{5}{6}$
11. $\frac{9}{10} \div 3$
12. $\frac{12}{13} \div \frac{1}{4}$

Solve each equation. Write the solution in simplest form.

13. $8 \div \frac{1}{2} = a$
14. $x = \frac{3}{5} \div \frac{9}{10}$
15. $\frac{5}{9} \div \frac{5}{6} = w$
16. $m = \frac{11}{12} \div 6$

7-5 Practice

Dividing Fractions

Find the reciprocal of each number.

1. $\frac{1}{4}$

2. $\frac{5}{6}$

3. 7

4. $\frac{8}{15}$

5. 12

6. $\frac{9}{14}$

7. $\frac{3}{11}$

8. 6

Find each quotient. Write in simplest form.

9. $\frac{5}{6} \div \frac{1}{3}$

10. $\frac{3}{4} \div \frac{5}{8}$

11. $\frac{1}{2} \div \frac{3}{5}$

12. $8 \div \frac{4}{5}$

13. $\frac{1}{6} \div \frac{2}{9}$

14. $\frac{9}{10} \div \frac{1}{4}$

15. $\frac{3}{8} \div 9$

16. $\frac{8}{9} \div \frac{2}{3}$

17. $\frac{2}{5} \div \frac{4}{7}$

18. $15 \div \frac{5}{9}$

19. $\frac{7}{8} \div \frac{7}{10}$

20. $\frac{1}{9} \div \frac{5}{12}$

21. $\frac{1}{5} \div \frac{7}{20}$

22. $\frac{5}{7} \div 10$

23. $\frac{7}{9} \div \frac{1}{7}$

Solve each equation. Write the solution in simplest form.

24. $j = \frac{6}{7} \div \frac{3}{14}$

25. $\frac{4}{9} \div \frac{14}{15} = b$

26. $\frac{9}{16} \div \frac{3}{4} = s$

27. $m = \frac{3}{5} \div \frac{9}{20}$

28. $\frac{7}{12} \div \frac{5}{6} = a$

29. $p = \frac{3}{8} \div \frac{9}{10}$

Evaluate each expression.

30. $y \div z$, if $y = \frac{4}{5}$ and $z = \frac{2}{3}$

31. $c \div d$, if $c = 14$ and $d = \frac{7}{8}$

32. $a \div b$, if $a = \frac{2}{9}$ and $b = \frac{1}{3}$

33. $e \div f$, if $e = 18$ and $f = \frac{3}{4}$

Mathematics: Applications and Connections, Course 1

Study Guide

Dividing Mixed Numbers

To divide mixed numbers, express each mixed number as an improper fraction. Then divide as with fractions.

Example Solve $m = 2\frac{5}{8} \div 1\frac{3}{4}$.   *Estimate: $2 \div 2 = 1$*

$m = \frac{21}{8} \div \frac{7}{4}$ *Express each mixed number as an improper fraction.*

$m = \frac{\overset{3}{\cancel{21}}}{\underset{2}{\cancel{8}}} \times \frac{\overset{1}{\cancel{4}}}{\underset{1}{\cancel{7}}}$ *Multiply by the reciprocal.*
Divide 21 and 7 by the GCF, 7.
Divide 4 and 8 by the GCF, 4.

$m = \frac{3}{2} \times \frac{1}{1}$ *Simplify.*

$m = \frac{3}{2}$ or $1\frac{1}{2}$ *Compare with your estimate.*

Find each quotient. Write in simplest form.

1. $2\frac{1}{2} \div \frac{4}{5}$

2. $1\frac{2}{3} \div 1\frac{1}{4}$

3. $5 \div 1\frac{3}{7}$

4. $2\frac{1}{3} \div \frac{7}{9}$

5. $5\frac{2}{5} \div \frac{9}{10}$

6. $7\frac{1}{2} \div 1\frac{2}{3}$

Solve each equation. Write the solution in simplest form.

7. $n = 10\frac{1}{2} \div \frac{7}{10}$

8. $3\frac{3}{5} \div 10 = p$

9. $r = 6\frac{3}{5} \div 2\frac{1}{5}$

10. $15 \div 3\frac{1}{3} = t$

11. $6\frac{2}{5} \div 1\frac{3}{5} = c$

12. $h = 2\frac{1}{12} \div 5$

13. $m = 18 \div \frac{9}{11}$

14. $r = 4\frac{4}{5} \div \frac{8}{15}$

15. $6\frac{3}{4} \div 1\frac{1}{8} = k$

7-6 Practice

Dividing Mixed Numbers

Write each mixed number as an improper fraction. Then write its reciprocal.

1. $8\frac{3}{4}$

2. $9\frac{6}{7}$

3. $7\frac{5}{6}$

4. $3\frac{5}{12}$

5. $1\frac{7}{16}$

6. $6\frac{7}{8}$

Find each quotient. Write in simplest form.

7. $4 \div 2\frac{2}{5}$

8. $3\frac{1}{4} \div 1\frac{3}{8}$

9. $\frac{8}{9} \div 5\frac{1}{3}$

10. $2\frac{1}{2} \div 4\frac{2}{7}$

11. $3\frac{1}{9} \div 7$

12. $6\frac{2}{3} \div 4\frac{4}{5}$

13. $2\frac{1}{7} \div \frac{3}{14}$

14. $3\frac{3}{5} \div 2\frac{4}{7}$

15. $9 \div 3\frac{3}{7}$

16. $1\frac{2}{9} \div 1\frac{5}{6}$

17. $\frac{7}{10} \div 2\frac{5}{8}$

18. $3\frac{1}{5} \div 1\frac{7}{9}$

19. $1\frac{3}{4} \div 14$

20. $2\frac{2}{15} \div 3\frac{5}{9}$

21. $2\frac{1}{10} \div \frac{7}{8}$

22. $6\frac{3}{4} \div 1\frac{7}{20}$

23. $18 \div 1\frac{1}{8}$

24. $4\frac{1}{6} \div 1\frac{3}{7}$

Solve each equation. Write the solution in simplest form.

25. $\frac{5}{12} \div 2\frac{1}{2} = p$

26. $s = 2\frac{2}{3} \div 1\frac{5}{6}$

27. $a = 1\frac{4}{5} \div 6$

28. $k = 2\frac{2}{5} \div 1\frac{7}{9}$

29. $1\frac{1}{6} \div \frac{5}{18} = d$

30. $1\frac{3}{5} \div 3\frac{4}{7} = x$

Evaluate each expression.

31. $f \div g$, if $f = 5\frac{1}{4}$ and $g = 1\frac{5}{9}$

32. $w \div z$, if $w = 9$ and $x = 2\frac{1}{7}$

33. $t \div v$, if $t = 6\frac{1}{2}$ and $v = 1\frac{7}{8}$

34. $j \div k$, if $j = 12$ and $k = 2\frac{4}{5}$

7-7 Study Guide

Integration: Measurement
Changing Customary Units

Customary Units	
Weight	**Liquid Capacity**
1 pound (lb) = 16 ounces (oz) 1 ton (T) = 2,000 pounds	1 cup (c) = 8 fluid ounces (fl oz) 1 pint (pt) = 2 cups 1 quart (qt) = 2 pints 1 gallon (gal) = 4 quarts

To change from larger units to smaller units, multiply.

Example 1 $4\frac{1}{2}$ lb = _____ oz   *Think: Each pound equals 16 ounces.*

$4\frac{1}{2} \times 16 = 72$ *Multiply to change from pounds to ounces.*

$4\frac{1}{2}$ lb = 72 oz

To change from smaller units to larger units, divide.

Example 2   **700 qt =** _____ **gal**   *Think: It takes 4 quarts to make 1 gallon.*

$700 \div 4 = 175$ *Divide to change from quarts to gallons.*

700 qt = 175 gal

Complete.

1. 2 lb = _____ oz   **2.** 3 T = _____ lb   **3.** 5 c = _____ fl oz

4. 1.5 lb = _____ oz   **5.** 10 pt = _____ qt   **6.** 12 qt = _____ gal

7. 12 fl oz = _____ c   **8.** 24 oz = _____ lb   **9.** 7 c = _____ fl oz

10. 2.5 pt = _____ c   **11.** 1.5 gal = _____ qt   **12.** 3.5 qt = _____ pt

13. 48 oz = _____ lb   **14.** 8,000 lb = _____ T   **15.** 2 pt = _____ c

7-7 Practice

Integration: Measurement
Changing Customary Units

Complete.

1. 6 pt = _____ c

2. 20 qt = _____ gal

3. 64 oz = _____ lb

4. 7 c = _____ fl oz

5. 12 qt = _____ pt

6. 12,000 lb = _____ T

7. 8 gal = _____ qt

8. 9 lb = _____ oz

9. 72 fl oz = _____ c

10. 4 T = _____ lb

11. 14 c = _____ pt

12. 26 pt = _____ qt

13. 10 qt = _____ gal

14. 2 T = _____ oz

15. 18 pt = _____ c

16. $3\frac{1}{2}$ c = _____ fl oz

17. 128 c = _____ gal

18. 96 oz = _____ lb

19. 21 qt = _____ pt

20. 750 lb = _____ T

21. 3 qt = _____ fl oz

22. 15 pt = _____ qt

23. 7 T = _____ lb

24. 2 gal = _____ c

25. 19 c = _____ pt

26. 4 qt = _____ c

27. $5\frac{1}{4}$ lb = _____ oz

28. 6 gal = _____ qt

29. 104 fl oz = _____ c

30. 64 pt = _____ gal

7-8 Study Guide

Integration: Patterns and Functions
Sequences

A **sequence** is a set of numbers that follows a pattern or rule.

Examples

1   3, 9, 27, 81, 243
×3 ×3 ×3 ×3

Each number in the sequence is multiplied by 3.

2   The next number in the sequence is 243 × 3 or 729.

400, 200, 100, 50
×$\frac{1}{2}$ ×$\frac{1}{2}$ ×$\frac{1}{2}$

Each number in the sequence is multiplied by $\frac{1}{2}$.

3   The next number in the sequence is 50 × $\frac{1}{2}$ or 25.

72, 68, 64, 60, 56
−4 −4 −4 −4

4 is subtracted from each number in the sequence.

4   The next number in the sequence is 56 − 4 or 52.

52.5, 55, 57.5, 60, 62.5
+2.5 +2.5 +2.5 +2.5

2.5 is added to each number in the sequence.

The next number in the sequence is 62.5 + 2.5 or 65.

Find the next two numbers in each sequence.

1. 72, 77, 82, 87, . . .

2. 3, 6, 12, 24, . . .

3. 100, 92, 84, 76, . . .

4. 400, 40, 4, 0.4, . . .

5. $7\frac{1}{2}$, 7, $6\frac{1}{2}$, 6, . . .

6. $\frac{1}{5}$, $\frac{1}{10}$, $\frac{1}{20}$, $\frac{1}{40}$, . . .

Find the missing number in each sequence.

7. 4, 12, _____, 108, . . .

8. 8,000, _____, 2,000, 1,000, . . .

9. _____, 13.3, 12.6, 11.9, . . .

10. $10\frac{1}{2}$, _____, $11\frac{1}{2}$, 12, . . .

Mathematics: Applications and Connections, Course 1

7-8 Practice

Integration: Patterns and Functions
Sequences

Find the next two numbers in each sequence.

1. 2, 8, 14, 20, . . .

2. 31, 27, 23, 19, . . .

3. $\frac{1}{3}$, 1, 3, 9, . . .

4. 108, 36, 12, 4, . . .

5. 43, 38, 33, 28, . . .

6. 1.2, 2.4, 3.6, 4.8, . . .

7. 3, 6, 12, 24, . . .

8. 63, 56, 49, 42, . . .

9. 1, $1\frac{2}{3}$, $2\frac{1}{3}$, 3, . . .

10. 4, 12, 36, 108, . . .

11. 81, 72, 63, 54, . . .

12. 6, 11, 16, 21, . . .

13. 5, 20, 35, 50, . . .

14. 27, 22.5, 18, 13.5, . . .

15. 18, 36, 54, 72, . . .

16. 1, 5, 25, 125, . . .

Find the missing number in each sequence.

17. $\frac{1}{6}$, $\frac{1}{3}$, _____, $\frac{2}{3}$, . . .

18. 54, _____, 42, 36, . . .

19. _____, 11, 14, 17, . . .

20. 1.7, 3.4, 5.1, _____, . . .

21. . . . , 200, 100, _____, 25

22. $1\frac{3}{4}$, $3\frac{1}{2}$, _____, 7, . . .

23. _____, 12, 48, 192, . . .

24. . . . , 91, 78, _____, 52, . . .

25. 9, _____, 23, 30, . . .

26. 0.4, 0.8, 1.2, _____, . . .

27. . . . , $\frac{3}{5}$, _____, $1\frac{4}{5}$, $2\frac{2}{5}$

28. 30, 26, _____, 18, . . .

29. _____, 7.5, 22.5, 67.5, . . .

30. . . . , 0.004, 0.04, _____, 4

8-1 Study Guide

Ratios and Rates

You can compare numbers with the same unit of measure using a **ratio**. A ratio is a comparison of two numbers by division.

Example 1 There were 52 animals in the ugly pet show. 24 of them were dogs.

The ratio can be written as follows.

24 to 52   24:52   24 out of 52 $\frac{24}{52}$

24 and 52 have 4 as a common factor.
The ratio can be simplified as $\frac{6}{13}$.

$$\frac{24}{52} = \frac{6}{13}$$
$\div 4$ $\div 4$

A **rate** is a ratio that compares two different units of measure.

Example 2 Julie entered 40 pages of data into her computer in 5 hours.

Compare the number of pages to the number of hours.

$$\frac{40 \text{ pages}}{5 \text{ hours}} = \frac{8 \text{ pages}}{1 \text{ hour}}$$
$\div 5$ $\div 5$

Simplify to find the unit rate, the number of pages entered in 1 hour.

Julie entered data at the rate of 8 pages per hour.

Express each ratio as a fraction in simplest form.

1. 12 out of 25 people

2. 8 out of 10 bicycles

3. 9 wins in 12 games

4. 14 station wagons out of 40 vehicles

5. 10 out of 15 days

6. 18 baseball caps out of 36 hats

Express each ratio as a rate.

7. $18 for 6 tickets

8. 30 km in 6 hours

9. $42 for 7 books

10. $28 for 4 hours

Use the letters in the word "WORLDWIDE." Write the ratios comparing the numbers of letters in simplest form.

11. W to D

12. R to W

13. vowels to consonants

8-1 Practice

Ratios and Rates

Write each ratio in three different ways.

1. 11 out of 32 students have blue eyes.

2. 4 out of 7 vacation days were sunny.

3. 19 out of 20 thank-you notes were written.

4. 8 out of 45 jelly beans are black.

Express each ratio as a fraction in simplest form.

5. read 75 pages out of 90

6. 8 blueberry muffins in $1\frac{1}{2}$ dozen muffins

7. 4 aces in a deck of 52 playing cards

8. 36 black keys out of 88 piano keys

9. 9 caramels in a box of 48 chocolates

10. score 24 points out of 72 points in basketball

11. 325 full-sized cars out of 500 cars

12. 42 pairs of hi-tops out of 56 pairs of sneakers

13. 16 male teachers out of 40 teachers

14. 35 sopranos in an 84-member chorus

15. exercise 45 out of 63 minutes

16. 18 goldfish in a tank of 48 fish

Express each ratio as a rate.

17. 105 words typed in 3 minutes

18. 10 pounds lost in 4 weeks

19. 2,800 miles driven in 7 days

20. $38 for 2 sweaters

21. $375 saved in 5 years

22. 72 hits in 24 baseball games

Using the letters of the phrase "STATE OF MASSACHUSETTS," write the ratios comparing the numbers of letters.

23. A to T

24. M to E

25. E to S

26. F to A

27. vowels to consonants

28. T to S

29. H to T

30. E to A

8-2 Study Guide

Solving Proportions

A **proportion** is an equation that shows that two ratios are equivalent. To determine if a pair of ratios form a proportion, you can find their cross products. If the cross products are equal, then the ratios form a proportion.

Examples   **1** Is $\frac{5}{6} = \frac{12}{18}$?

$$\frac{5}{6} \gtrless \frac{12}{18}$$

$5 \times 18 \stackrel{?}{=} 6 \times 12$ *Write cross products.*
$90 \neq 72$ *Multiply.*

$$\frac{5}{6} \neq \frac{12}{18}$$

$\frac{5}{6}$ and $\frac{12}{18}$ do not form a proportion.

2 Is $\frac{12}{32} = \frac{3}{8}$?

$$\frac{12}{32} \gtrless \frac{3}{8}$$

$12 \times 8 \stackrel{?}{=} 32 \times 3$
$96 = 96$

$$\frac{12}{32} = \frac{3}{8}$$

$\frac{12}{32}$ and $\frac{3}{8}$ form a proportion.

You can use cross products to solve proportions.

Example 3   Solve $\frac{15}{24} = \frac{10}{c}$.

$15 \times c = 24 \times 10$	*Write cross products.*
$15c = 240$	*Multiply.*
$15c \div 15 = 240 \div 15$	*Divide.*
$c = 16$	

Use cross products to determine whether each pair of ratios forms a proportion.

1. $\frac{8}{12}, \frac{5}{10}$

2. $\frac{12}{30}, \frac{2}{5}$

3. $\frac{8}{24}, \frac{6}{18}$

4. $\frac{10}{24}, \frac{5}{8}$

5. $\frac{12}{15}, \frac{3}{4}$

6. $\frac{4}{5}, \frac{20}{25}$

Solve each proportion.

7. $\frac{8}{15} = \frac{m}{45}$

8. $\frac{9}{12} = \frac{6}{c}$

9. $\frac{5}{p} = \frac{3}{9}$

10. $\frac{v}{21} = \frac{4}{6}$

11. $\frac{14}{8} = \frac{x}{4}$

12. $\frac{9}{r} = \frac{27}{30}$

13. $\frac{10}{4} = \frac{m}{20}$

14. $\frac{f}{4} = \frac{4}{16}$

15. $\frac{1}{c} = \frac{12}{24}$

8-2 Practice

Solving Proportions

Use cross products to determine whether each pair of ratios forms a proportion.

1. $\frac{4}{9}, \frac{12}{27}$

2. $\frac{5}{6}, \frac{25}{36}$

3. $\frac{5}{12}, \frac{7}{18}$

4. $\frac{2}{7}, \frac{16}{56}$

5. $\frac{3}{14}, \frac{9}{42}$

6. $\frac{3}{4}, \frac{18}{24}$

7. $\frac{20}{32}, \frac{5}{8}$

8. $\frac{21}{40}, \frac{3}{5}$

Solve each proportion.

9. $\frac{6}{7} = \frac{a}{56}$

10. $\frac{27}{x} = \frac{3}{8}$

11. $\frac{2}{3} = \frac{34}{p}$

12. $\frac{s}{54} = \frac{7}{9}$

13. $\frac{4}{5} = \frac{c}{75}$

14. $\frac{r}{48} = \frac{7}{12}$

15. $\frac{8}{15} = \frac{24}{t}$

16. $\frac{36}{m} = \frac{2}{6}$

17. $\frac{3}{11} = \frac{21}{b}$

18. $\frac{1}{4} = \frac{y}{52}$

19. $\frac{30}{g} = \frac{5}{16}$

20. $\frac{n}{60} = \frac{7}{12}$

21. $\frac{d}{12} = \frac{6}{8}$

22. $\frac{7}{35} = \frac{3}{k}$

23. $\frac{8}{f} = \frac{15}{60}$

24. $\frac{14}{42} = \frac{z}{18}$

25. $\frac{6}{9} = \frac{b}{15}$

26. $\frac{4}{w} = \frac{5}{25}$

27. $\frac{4}{6} = \frac{14}{g}$

28. $\frac{6}{10} = \frac{j}{45}$

State a proportion for each of the following. Then solve the proportion.

29. If you spend 1.5 hours per day doing homework, how many hours would you spend doing homework in 8 days?

30. If you can buy 2 videos for $35.96, how much will 5 videos cost?

8-3 Study Guide

Integration: Geometry
Scale Drawings

A scale drawing and the object it illustrates are **similar figures**.
The lengths on the drawing are proportional to the actual lengths.

Example This scale drawing of a grand piano (top view) has a scale of $\frac{1}{4}$ inch = 1 foot. The front of the grand piano measures $1\frac{1}{4}$ inches on the drawing. What is the actual length?

Write a proportion to find the actual length.

length in drawing ⟶ $\dfrac{\frac{1}{4} \text{ inch}}{1 \text{ foot}}$ = $\dfrac{1\frac{1}{4} \text{ inch}}{\ell}$ ⟵ length in drawing

actual length ⟶ ⟵ actual length

$\frac{1}{4} \times \ell = 1 \times 1\frac{1}{4}$ *Find the cross products.*

$\frac{1}{4} \times \ell = 1\frac{1}{4}$ *Multiply.*

$4\left(\frac{1}{4}\right)\ell = 4\left(1\frac{1}{4}\right)$ *Multiply each side by 4.*

$\ell = 5$ *Since $4 \times \frac{1}{4} = 1$, $\ell = 5$.*

The bedroom floor plan at the right has a scale of $\frac{1}{4}$ inch = 1 foot. Use a ruler to measure each object to the nearest $\frac{1}{4}$ inch. Then find the actual measurements.

	Scale Drawing		Actual	
	Length	**Width**	**Length**	**Width**
Room				
Table				
Dresser				
Bookcase				
Desk				
Bed				

8-3 Practice

Integration: Geometry
Scale Drawings

1. The drawing of the rocket has a scale of $\frac{1}{2}$ inch = 12 feet. Use a ruler to measure the drawing to the nearest $\frac{1}{4}$ inch and compute the actual measurements of the rocket.

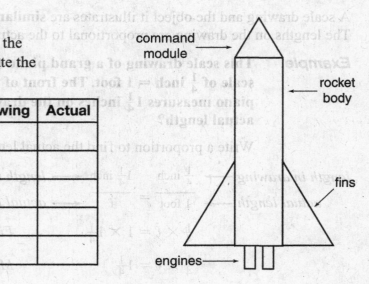

command module

rocket body

fins

engines

	Drawing	Actual
a. height of command module		
b. height of rocket body		
c. height of engines		
d. width of rocket body and fins		
e. total length		

2. The proposed plan for a new amusement park at the right has a scale of 1 centimeter = 5 feet. Use a centimeter ruler to measure each section of the park to the nearest centimeter. Then compute the actual measurements.

rides

arcades

snack bar

picnic area

zoo

roller coaster

Attraction	Drawing		Actual	
	length (cm)	width (cm)	length (ft)	width (ft)
a. rides				
b. roller coaster				
c. snack bar				
d. picnic area				
e. arcades				
f. zoo				

8-4 Study Guide

Percents and Fractions

A **percent** is a ratio that compares a number to 100.

To express a percent as a fraction, write it as a fraction with a denominator of 100 and then simplify.

Example 1 $48\% = \frac{48}{100}$

$= \frac{12}{25}$ *Divide the numerator and denominator by 4.*

To express a fraction as a percent, first set up a proportion. Then solve the proportion using cross products.

Example 2 Express $\frac{13}{20}$ as a percent.

$\frac{13}{20} = \frac{k}{100}$ *Set up a proportion.*

$13 \times 100 = 20 \times k$ *Find the cross products.*

$1{,}300 = 20k$

$1{,}300 \div 20 = 20k \div 20$ *Divide each side by 20.*

$65 = k$

$\frac{13}{20} = \frac{65}{100}$ or 65%

Express each percent as a fraction in simplest form.

1. 60%   2. 75%   3. 50%   4. 99%

5. 12%   6. 45%   7. 54%   8. 5%

Express each shaded section as a fraction and as a percent.

9.    10.   11.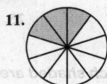

Express each fraction as a percent.

12. $\frac{17}{100}$ 13. $\frac{4}{5}$ 14. $\frac{1}{4}$ 15. $\frac{8}{20}$

16. $\frac{1}{50}$ 17. $\frac{7}{10}$ 18. $\frac{6}{25}$ 19. $\frac{1}{10}$

Practice

Name_____ Date_____

Percents and Fractions

Express each percent as a fraction in simplest form.

1. 24%   2. 98%   3. 35%   4. 7%

5. 120%   6. 54%   7. 76%   8. 5%

9. 82%   10. 130%   11. 19%   12. 175%

Use a 10 × 10 grid to shade the amount stated in each fraction. Then express each fraction as a percent.

13. $\frac{1}{10}$ 14. $\frac{1}{20}$ 15. $\frac{1}{50}$

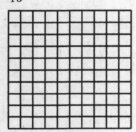 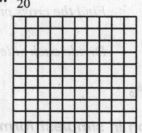 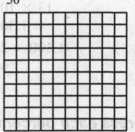

Express each fraction as a percent.

16. $\frac{47}{100}$ 17. $\frac{8}{25}$ 18. $\frac{9}{10}$ 19. $\frac{13}{50}$

20. $\frac{11}{20}$ 21. $\frac{7}{5}$ 22. $\frac{3}{100}$ 23. $\frac{21}{25}$

24. $\frac{7}{10}$ 25. $\frac{3}{20}$ 26. $\frac{31}{50}$ 27. $\frac{5}{4}$

Express each shaded area as a fraction and as a percent.

28.   29.   30.

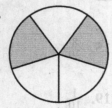 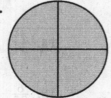 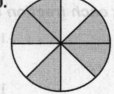

Mathematics: Applications and Connections, Course 1

8-5 Study Guide

Percents and Decimals

To express a percent as a decimal, rewrite the percent as a fraction with a denominator of 100. Then express the fraction as a decimal.

Examples   Express each percent as a decimal.

$$1 \quad 29\% = \frac{29}{100} \qquad 2 \quad 0.7\% = \frac{0.7}{100} \qquad 3 \quad 145\% = \frac{145}{100}$$
$$= 0.29 \qquad\qquad\qquad = \frac{7}{1,000} \qquad\qquad = 1.45$$
$$= 0.007$$

To express a decimal as a percent, first express the decimal as a fraction with a denominator of 100. Then express the fraction as a percent.

Examples   Express each decimal as a percent.

$$4 \quad 0.09 = \frac{9}{100} \qquad 5 \quad 0.005 = \frac{5}{1,000} \qquad 6 \quad 1.8 = \frac{18}{10}$$
$$= 9\% \qquad\qquad\qquad = \frac{0.5}{100} \qquad\qquad = \frac{180}{100}$$
$$= 0.5\% \qquad\qquad = 180\%$$

Express each percent as a decimal.

1. 97%   2. 6%   3. 15%   4. 0.8%

5. 362%   6. 62.3%   7. 3.5%   8. 708%

Express each decimal as a percent.

9. 0.66   10. 0.08   11. 0.75   12. 0.001

13. 1.19   14. 0.72   15. 0.136   16. 4.02

8-5 Practice

8-5 Study Guide

Percents and Decimals

Express each percent as a decimal.

1. 96%   2. 52%   3. 60%   4. 42.5%

5. 8%   6. 250%   7. 0.1%   8. 7.4%

9. 34.5%   10. 19%   11. 6.2%   12. 19%

13. 0.5%   14. 37%   15. 90%   16. 4%

Express each decimal as a percent.

17. 0.18   18. 0.36   19. 0.09   20. 0.2

21. 0.625   22. 0.007   23. 1.4   24. 0.093

25. 0.8   26. 0.54   27. 3.75   28. 0.02

29. 0.258   30. 0.016   31. 0.49   32. 0.003

Mathematics: Applications and Connections, Course 1

8-6 Study Guide

Estimating with Percents

You can use rounding to estimate with percents.

Examples

1 **Estimate 75% of 788.**

$75\% = \frac{3}{4}$ *Express 75% as a fraction.*

Round 788 to 800.

$\frac{3}{4} \times 800 = 600$ *Multiply.*

75% of 788 is about 600.

2 **Estimate 65% of 90.**

65% is close to $66\frac{2}{3}\%$ or $\frac{2}{3}$.   *Express 65% as a compatible fraction.*

$\frac{2}{3} \times 90 = 60$ *Estimate the product using the fraction.*

65% of 90 is about 60.

3 **Estimate 48% of 192.**

48% is close to 50% or $\frac{1}{2}$.   *Express 48% as a fraction.*

Round 192 to 200.

$\frac{1}{2} \times 200 = 100$ *Multiply.*

48% of 192 is about 100.

Estimate each percent.

1. 31% of 150

2. 50% of 27

3. 79% of 102

4. 9% of 450

5. 23% of 60

6. 30% of 96

7. 53% of 82

8. 21% of 200

9. 68% of 89

10. 10.7% of 160

11. 79% of 21

12. 0.6% of 201

8-6 Practice

Estimating with Percents

Estimate each percent.

1. 18% of 35

2. 26% of 48

3. 79% of 40

4. 52% of 110

5. 34% of 120

6. 68% of 82

7. 43% of 59

8. 77% of 160

9. 28% of 142

10. 66% of 91

11. 59% of 25

12. 11% of 178

13. 4.8% of 40

14. 92% of 29

15. 49% of 56

16. 62% of 10

17. 32% of 270

18. 24% of 203

19. 9% of 98

20. 81% of 45

21. 69% of 122

22. 7.7% of 50

23. 38% of 109

24. 89% of 61

25. 76% of 43

26. 21% of 95

27. 67% of 240

Use the chart at the right to estimate each percent.

28. percent of first twenty prime numbers from 2 through 71 that contain the digit 4

29. percent of first twenty prime numbers from 11 through 71 whose sum of the digits is an even number

30. percent of first twenty prime numbers from 2 through 71 that contain the digit 7

First Twenty Prime Numbers			
2	3	5	7
11	13	17	19
23	29	31	37
41	43	47	53
59	61	67	71

Mathematics: Applications and Connections, Course 1

8-7 Study Guide

Percent of a Number

One way to find the percent of a number is to change the percent to a fraction and then multiply.

Example 1 **Find 40% of 90 by changing the percent to a fraction.**

$40\% = \frac{40}{100}$ or $\frac{2}{5}$ *Change the percent to a fraction.*

$90 \times \frac{2}{5} = 36$ *Multiply the number by the fraction.*

40% of 90 is 36.

Another way to find the percent of a number is to change the percent to a decimal and then multiply.

Example 2 **Find 8% of 68 by changing the percent to a decimal.**

$8\% = 0.08$ *Change the percent to a decimal.*

$0.08 \times 68 = 5.44$ *Multiply the number by the decimal.*

8% of 68 is 5.44.

Find the percent of each number.

1. 12% of 140   **2.** 25% of 164   **3.** 65% of 125

4. 77% of 90   **5.** 16% of 48   **6.** 55% of 96

7. 105% of 62   **8.** 340% of 91   **9.** 0.5% of 180

10. 2% of 84   **11.** 200% of 13   **12.** 5% of 80

Percent of a Number

Find the percent of each number.

1. 75% of 52

2. 40% of 65

3. 15% of 80

4. 30% of 24

5. 62.5% of 96

6. 9% of 20

7. 28% of 75

8. 95% of 60

9. 70% of 15

10. 12% of 300

11. 85% of 48

12. 125% of 16

13. 0.6% of 5

14. 36% of 175

15. 48% of 50

16. 160% of 90

17. 65% of 120

18. 87.5% of 56

19. 5% of 85

20. 90% of 18

21. 0.4% of 150

22. 120% of 70

23. 37.5% of 104

24. 52% of 25

25. 80% of 40

26. 45% of 200

27. 2.5% of 4

28. 34% of 30

29. 55% of 95

30. 150% of 110

31. Gina read 35% of her 140-page book. How many pages did she read?

32. Larry delivered 75% of his 120 newspapers. How many papers did he deliver?

Mathematics: Applications and Connections, Course 1

9-1 Study Guide

Angles

To measure an angle, place the center of a protractor on the
vertex of the angle. Place the zero mark of the scale along one
side of the angle. Read the angle measure where the other side
of the angle crosses the scale.

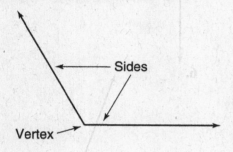

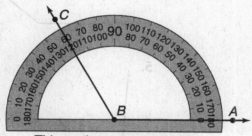

This angle measures 120°.

Angles can be classified according to their measure.

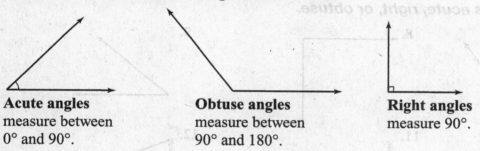

Acute angles
measure between
0° and 90°.

Obtuse angles
measure between
90° and 180°.

Right angles
measure 90°.

Use a protractor to find the measure of each angle.

1.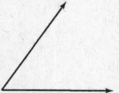

2.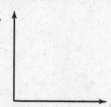

3.

Classify each angle as acute, right or obtuse.

4.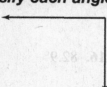

5.

6.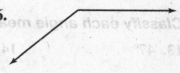

Classify each angle measure as acute, right, or obtuse.

7. 172°

8. 83°

9. 12°

Name _____ Date _____

Angles

Use a protractor to find the measure of each angle.

1.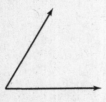

2.

3.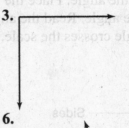

4.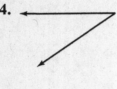

5.

6.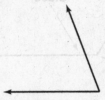

Classify each angle as acute, right, or obtuse.

7.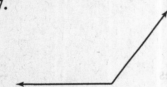

8.

9.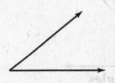

10.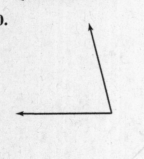

11.

12.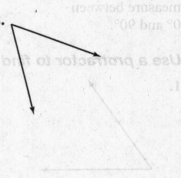

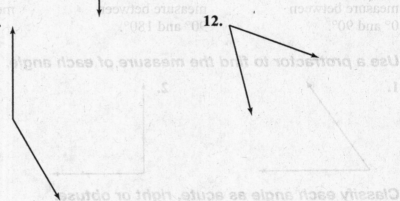

Classify each angle measure as acute, right, or obtuse.

13. 47°   14. 95°   15. 16°   16. 82.9°

17. 90°   18. 153°   19. 179°   20. 25°

9-2 Study Guide

Using Angle Measures

Use a protractor to draw an angle with a certain measure.

Example 1   **Draw a 75° angle.**

Draw one side.
Mark the vertex
and draw an arrow.

Place the protractor's
center mark on the vertex.
Place the zero mark of the
scale on the side. Find
75° and mark the point.

Draw the side
that connects the
vertex and the
pencil mark.

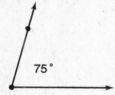

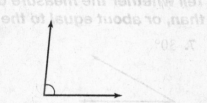

You can use the measures
of these angles to estimate
measures of other angles.

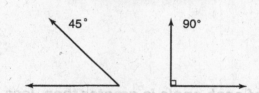

45°   90°   135°

Example 2   **Estimate the measure of the angle shown by using
multiples of a 45° angle.**
The angle shown is about the same as a 90° angle
wedge. So, the measure of the angle is about 90°.

Use a protractor and a straightedge to draw angles having the following measurements.

1. 35°   2. 110°   3. 15°   4. 175°

Tell whether the measure of each angle is greater than, less than, or about equal to the measurement given.

5. 140°   6. 87°   7. 25°   8. 7°

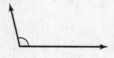      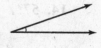

Practice

Study Guide

Name _____ **Date** _____

Using Angle Measures

Use a protractor and a straightedge to draw angles having the following measurements.

1. 110°

2. 12°

3. 64°

4. 159°

5. 48°

6. 123°

Tell whether the measure of each angle is greater than, less than, or about equal to the measurement given.

7. 30°

8. 175°

9. 85°

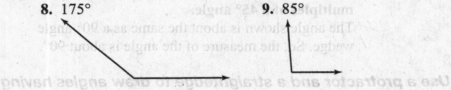

10. 98

11. 134

12. 72

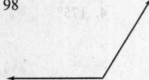

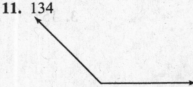

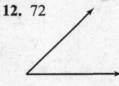

Without a protractor, draw your best estimate of an angle having each measurement given. Check your estimate with a protractor.

13. 100°

14. 57°

15. 168°

9-3 Study Guide

Constructing Bisectors

A **bisector** divides a line segment or an angle into two equal parts.

Examples   **1**   **Use a straightedge and a compass to bisect $\overline{AB}$.**

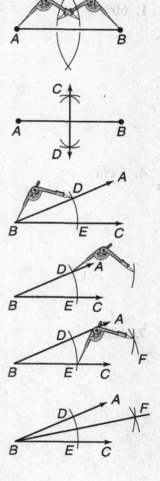

- Use the straightedge to draw $\overline{AB}$.

- Place the compass point at A. Set the compass for more than half the length of $\overline{AB}$. Draw two arcs as shown.

- With the same compass setting, place the compass point at B and draw two arcs as shown. These arcs should intersect the first arc at C and D.

- With a straightedge, draw $\overline{CD}$. $\overline{CD}$ bisects $\overline{AB}$.

2   **Use a straightedge and a compass to bisect $\angle ABC$.**

- Draw $\angle ABC$ using a straightedge.

- Place the compass point at B and draw an arc that intersects both sides of the angle. Label these points D and E.

- Place the compass point at D and draw an arc as shown.

- With the same compass setting, place the compass point at E and draw an arc that intersects the one drawn in the previous step. Label the intersection F.

- Using a straightedge, draw $\overrightarrow{BF}$. $\overrightarrow{BF}$ bisects $\angle ABC$.

Draw the angle or line segment with the given measurement. Then use a straightedge and a compass to bisect each angle or line segment.

1. 72°

2. 50 mm

3. 152°

4. $1\frac{1}{2}$ in.

5. 49°

6. 6 cm

9-3 Practice

Constructing Bisectors

**Draw the angle or line segment with the given measurement.
Then use a straightedge and a compass to bisect each angle
or line segment.**

1. 60°

2. 15 mm

3. 100°

4. 5 cm

5. 25°

6. 1 in.

7. 120°

8. 2 cm

9. 40°

10. 2.25 in.

11. 78°

12. 35 mm

Name _____ Date _____

9-4 Study Guide

Two-Dimensional Figures

A **quadrilateral** is a four-sided polygon. Quadrilaterals may be classified by looking at their sides and angles.

Square

All sides are congruent. All four angles are right angles.

Rectangle

Both pairs of opposite sides are parallel. All four angles are right angles.

Parallelogram

Both pairs of opposite sides are parallel.

A polygon is named according to the number of its sides.

Pentagon (5 sides)   Hexagon (6 sides)   Octagon (8 sides)   Decagon (10 sides)

Name each polygon.

1.

2.

3.

4.

Explain how each pair of figures is alike and how each pair is different.

5.

6.

Draw an example of each polygon. Mark any congruent sides, congruent angles, and right angles.

7. parallelogram   8. pentagon   9. rectangle   10. decagon

Mathematics: Applications and Connections, Course 1

9-4 Practice

Two-Dimensional Figures

Name each polygon.

1.

2.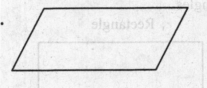

3.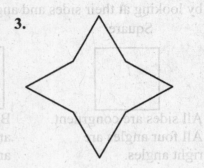

4.

5.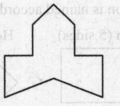

6.

Explain how each pair of figures is alike and how each pair is different.

7.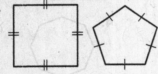

8.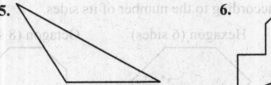

Draw an example of each polygon.

9. octagon

10. rectangle

11. decagon

9-5 Study Guide

Lines of Symmetry

If a figure can be folded in half so that the two halves match exactly, the figure has a **line of symmetry**.

Examples   one line of symmetry   two lines of symmetry   no lines of symmetry

Tell whether the dashed line is a line of symmetry. Write yes or no.

1.
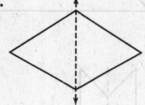

2.
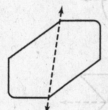

3.
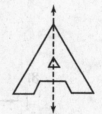

4.
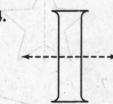

5.
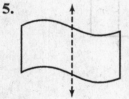

6.
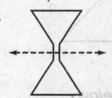

7.

8.

Draw all lines of symmetry.

9.
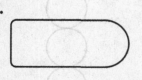

10.
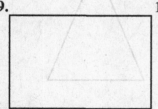

11.

12.

13.
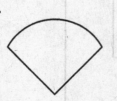

14.
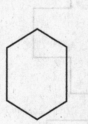

15.
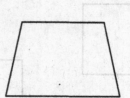

16.

Mathematics: Applications
and Connections, Course 1

9-5 Practice

Lines of Symmetry

Tell whether the dashed line is a line of symmetry. Write yes or no.

1.
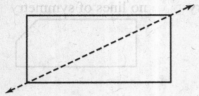

2.
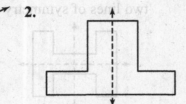

3.
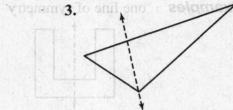

4.

5.
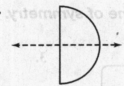

6.
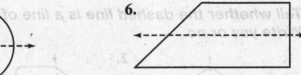

7.

8.
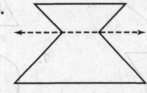

9.
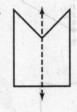

Draw all lines of symmetry for each figure below.

10.
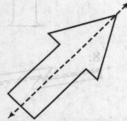

11.
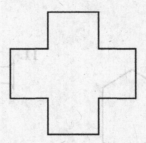

12.

13.
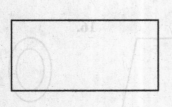

14.
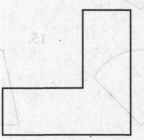

15.
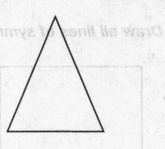

Mathematics: Applications and Connections, Course 1

9-6 Study Guide

Size and Shape

Figures that are the same size and shape are **congruent figures.**
The symbol ≅ means "is congruent to."

Figures that have the same shape but different size are **similar
figures.** The symbol ~ means "is similar to."

Examples Is each pair of polygons congruent, similar, or neither?

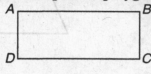
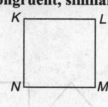

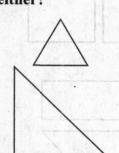

The rectangles are
the same size and
the same shape.
ABCD ≅ EFGH.

The shapes are
the same shape but
not the same size.
KLMN ~ PQRS.

The triangles are
neither the same size
nor the same shape.

**Tell whether each pair of polygons is congruent, similar, or
neither.**

1.

2.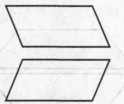

3.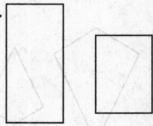

4.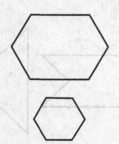

5.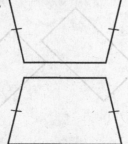

6.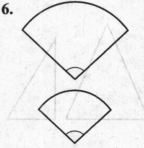

9-6

Practice

Size and Shape

Tell whether each pair of polygons is congruent, similar, or neither.

1.

2.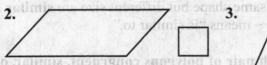

3.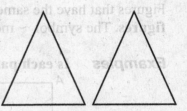

4.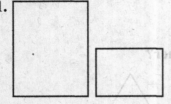

5.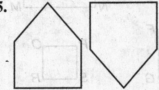

6.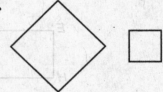

7.

8.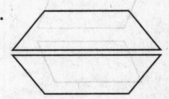

9.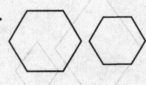

10.

11.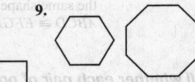

12.

13.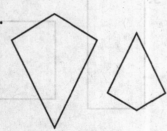

14.

15.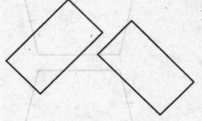

10-1 **Study Guide**

10-1 **Practice**

Area of Parallelograms

Area of Parallelograms

The area (A) of a parallelogram equals the product of
its base (b) and its height (h): $A = bh$.

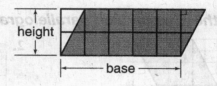

Example   **Find the area of the parallelogram.**

$A = bh$
$A = 7.4 \times 5.2$
$A = 38.48$

The area of the parallelogram is
38.48 square centimeters.

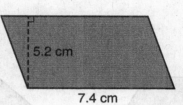

5.2 cm

7.4 cm

Find the area of each parallelogram to the nearest tenth.

1.
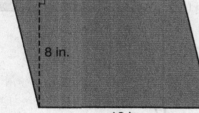
8 in.
12 in.

2.
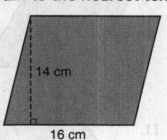
14 cm
16 cm

3.
9 ft
15 ft

4.
10.5 m
6.2 m

5.
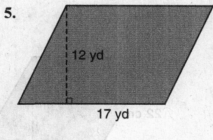
12 yd
17 yd

6.
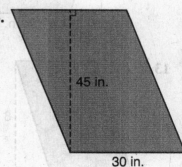
45 in.
30 in.

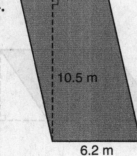
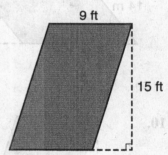

10-1 Practice

Area of Parallelograms

Find the area of each parallelogram.

1.
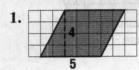
4
5

2.
6
3

3.
7 ft
9 ft

4.
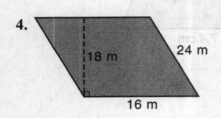
24 m
18 m
16 m

5.
3.6 cm
4.5 cm
13.5 cm

6.
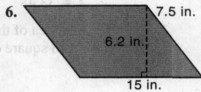
7.5 in.
6.2 in.
15 in.

7.
14 m
11 m
19 m

8.
21 cm
17 cm
23 cm

9.
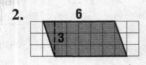
5½ ft
8 ft
24 ft

10.
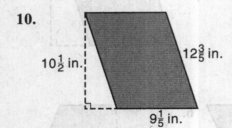
10½ in.
12⅗ in.
9⅕ in.

11.
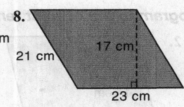
13 m
12 m
15.4 m

12.
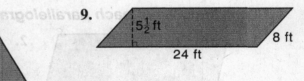
16 mm
7.2 mm
32 mm

13.
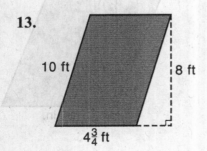
10 ft
8 ft
4¾ ft

14.
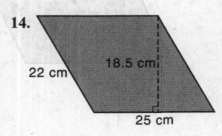
22 cm
18.5 cm
25 cm

15.
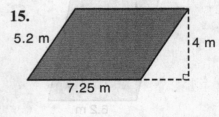
5.2 m
4 m
7.25 m

10-2 Study Guide

Area of Triangles

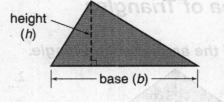

The area (*A*) of a triangle equals one half the product of its base (*b*) and height (*h*):
$A = \frac{1}{2}bh$.

Example   **Find the area of the triangle.**

$A = \frac{1}{2}bh$

$A = \frac{1}{2} \times 15 \times 26$

$A = 195$

The area of the triangle is 195 square feet.

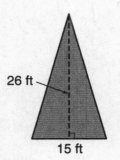

26 ft

15 ft

Find the area of each triangle. Round decimal answers to the nearest tenth.

1. base, 12 inches
   height, 7 inches

2. base, 20 cm
   height, 12 cm

3. base, 8 ft
   height, 24 ft

4.

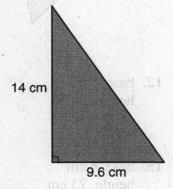

14 cm

9.6 cm

5.

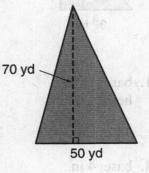

70 yd

50 yd

6.

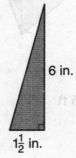

6 in.

$1\frac{1}{2}$ in.

7. Find the area of a triangle with a base of 40 miles and a height of 20 miles.

10-2 Practice

Area of Triangles

Find the area of each triangle.

1.

5 cm
14 cm

2.

18 in.
24 in.

3.

8.4 m
7 m

4.

3 ft
12 ft

5.

9 mm
26 mm

6.

10.5 m
17.2 m

7.

19 yd
13 yd

8.

$4\frac{2}{5}$ in.
$3\frac{3}{5}$ in.

9.

6.8 cm
15.5 cm

10. base, 9 ft
height, 16 ft

11. base, 18 m
height, 6 m

12. base, 5 yd
height, 19 yd

13. base, 15 mm
height, 14 mm

14. base, 4 in.
height, 11 in.

15. base, 8 cm
height, 23 cm

16. Find the area of a triangle with a base of 42 kilometers and a height of 25 kilometers.

10-

Study Guide

Area of Circles

The area (A) of a circle equals the product of and the square of π the radius (r): $A = \pi r^2$.

Example   **Find the area of the circle.**

The radius is one-half of the diameter.
The radius of the circle is $\frac{1}{2}(14)$ or 7 meters.

$A = \pi r^2$
$A \approx 3.14 \times 7^2$ *Use 3.14 for π.*
$A \approx 3.14 \times 49$ *$7^2 = 7 \times 7$ or 49*
$A \approx 153.86$

The area of the circle is about 153.86 square meters.

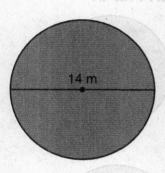

14 m

Find the area of each circle to the nearest tenth. Use 3.14 for π.

1.

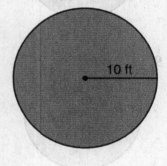

10 ft

2.

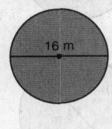

16 m

3.

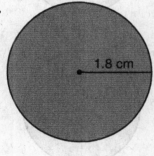

1.8 cm

4. radius, 12 inches

5. radius, 4 meters

6. radius, 9 feet

7. diameter, 10 centimeters

8. diameter, 22 inches

9. diameter, 12 yards

Mathematics: Applications and Connections, Course 1

10-3 Practice

10-3 Study Guide

Area of Circles

**Find the area of each circle to the nearest tenth.
Use 3.14 for π.**

1.
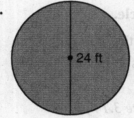
5 cm

2.
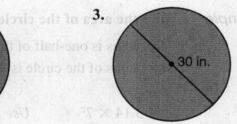
24 ft

3.
30 in.

4.
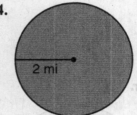
2 mi

5.
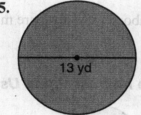
13 yd

6.
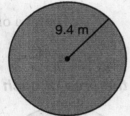
9.4 m

7.
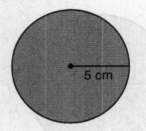
14 ft

8.
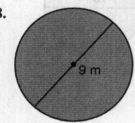
9 m

9.
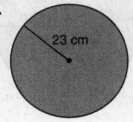
23 cm

10. diameter, 25 inches

11. radius, 13 yards

12. radius, 7.5 meters

13. diameter, $2\frac{1}{2}$ miles

14. radius, 30 feet

15. diameter, 34 meters

16. diameter, 11 miles

17. radius, 10.5 inches

18. diameter, 17 yards

19. Find the area of a circle if the radius is 24 meters.

20. What is the area of a circle with a diameter of 40 feet?

Study Guide

10-4 Practice
Three-Dimensional Figures
Name each figure.

Three Dimensional Figures

Pyramids are named by their bases.

 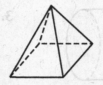

triangular pyramid   square pyramid

Prisms have two parallel congruent bases.

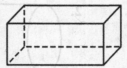 

rectangular prism   hexagonal prism

The polygons that form three-dimensional figures are called **faces**.
The faces intersect to form **edges**.
The edges intersect to form **vertices**.

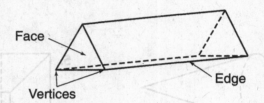

Face
Vertices
Edge

Some three-dimensional figures have curved surfaces.
They do not have polygons for faces.

 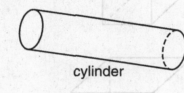

cone   cylinder   sphere

Name each figure.

1. 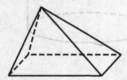   2.   3.   4.

Complete the chart by writing the number of faces, edges, and vertices for each figure.

Figure	Number of Faces	Number of Edges	Number of Vertices
10. triangular pyramid			
11. rectangular prism			
12. cylinder			
13. sphere			
14. square pyramid			
15. cone			
16. hexagonal pyramid			

State the number of faces, edges, and vertices in each figure.

5. triangular prism

6. rectangular pyramid

7. hexagonal prism

8. pentagonal pyramid

Name_____ Date_____

10-4 Practice

Three-Dimensional Figures

Name each figure.

1.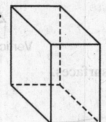

2.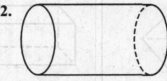

3.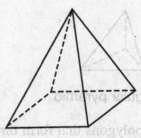

4.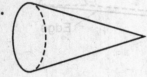

5.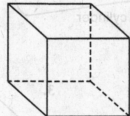

6.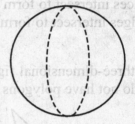

7.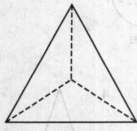

8.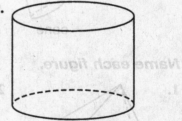

9.

Complete the chart by writing the number of faces, edges, and vertices for each figure.

	Figure	Number of Faces	Number of Edges	Number of Vertices
10.	triangular pyramid			
11.	rectangular prism			
12.	cylinder			
13.	sphere			
14.	square pyramid			
15.	cone			
16.	hexagonal pyramid			

Mathematics: Applications and Connections, Course 1

10-5 Study Guide

Volume of Rectangular Prisms

The volume (V) of a rectangular prism equals the product of its length (ℓ), its width (w), and its height (h): $V = \ell w h$.

Example **Find the volume of the rectangular prism.**

$V = \ell w h$
$V = 20 \times 15 \times 9$
$V = 2,700$

The volume of the rectangular prism is 2,700 m³.

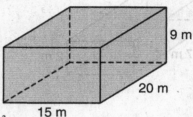

9 m
20 m
15 m

Find the volume of each rectangular prism.

1.

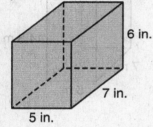

6 in.
7 in.
5 in.

2.

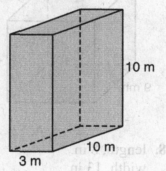

10 m
10 m
3 m

3.

4 ft
8 ft
12 ft

4.

9 cm
9 cm
9 cm

5.

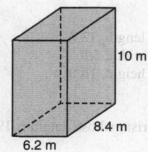

10 m
8.4 m
6.2 m

6.

3 yd
15 yd
3 yd

7. length, 11 ft
   width, 6 ft
   height, 15 ft

8. length, 9 mm
   width, 12 mm
   height, 20 mm

10-5 Practice

10-5 Study Guide

Volume of Rectangular Prisms

Find the volume of each rectangular prism.

1.
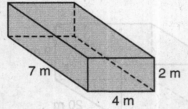
7 m   2 m   4 m

2.
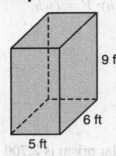
9 ft   6 ft   5 ft

3.
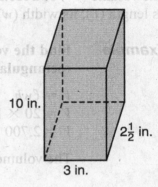
10 in. $2\frac{1}{2}$ in.   3 in.

4.
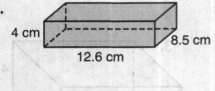
4 cm   8.5 cm   12.6 cm

5.
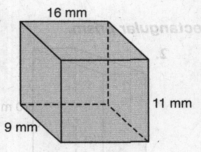
16 mm   11 mm   9 mm

6.
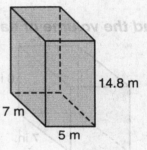
14.8 m   7 m   5 m

7. length, 21 mm
   width, 15 mm
   height, 18 mm

8. length, 8 in.
   width, 13 in.
   height, 17 in.

9. length, 11 cm
   width, 6.3 cm
   height, 3.6 cm

10. length, 4 m
    width, 19 m
    height, 10.6 m

11. length, 12 ft
    width, 5 ft
    height, 16 ft

12. length, 24 mm
    width, 3 mm
    height, 9 mm

13. Find the volume of a rectangular prism whose length is 19 feet, width is 15 feet, and height is 17 feet.

14. The town of Riverview provides a rectangular recycling bin for each household. The bin measures $20\frac{1}{2}$ inches by 12 inches by 16 inches. Find the volume of the recycling bin.

15. Janine keeps her jewelry in a jewelry box that measures 9 centimeters by 4.5 centimeters by 3 centimeters. What is the volume of her jewelry box?

10-6 Study Guide

Surface Area of Rectangular Prisms

The surface area of a rectangular prism is equal to the sum
of the areas of its faces.

Example   **Find the surface area of the rectangular prism.**

Find the area of each face.

front:	$6 \times 10 = 60$ cm²
back:	$6 \times 10 = 60$ cm²
top:	$8 \times 10 = 80$ cm²
bottom:	$8 \times 10 = 80$ cm²
right side:	$6 \times 8 = 48$ cm²
left side:	$6 \times 8 = 48$ cm²

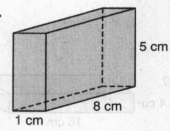

Add the areas: $60 + 60 + 80 + 80 + 48 + 48 = 376$.

The surface area of the rectangular prism is 376 cm².

Find the surface area of each rectangular prism.

1.
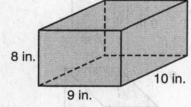
8 in. 10 in. 9 in.

2.
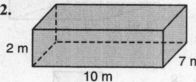
2 m 10 m 7 m

3.
5 cm 8 cm 1 cm

4.
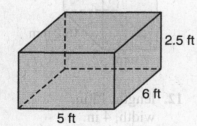
2.5 ft 6 ft 5 ft

5.
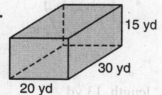
15 yd 30 yd 20 yd

6.
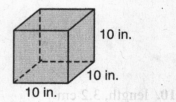
10 in. 10 in. 10 in.

7. length, 4 ft
   width, 6 ft
   height, 20 ft

8. length, 7 cm
   width, 8 cm
   height, 8 cm

10-6 Practice

Surface Area of Rectangular Prisms

Find the surface area of each rectangular prism.

1.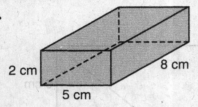
2 cm 8 cm
5 cm

2.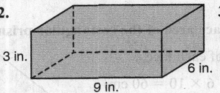
3 in. 6 in.
9 in.

3.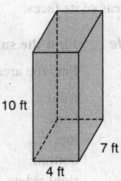
10 ft
7 ft
4 ft

4.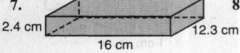
4 mm 8 mm
14 mm

5.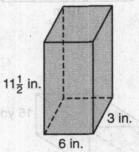
$7\frac{1}{2}$ yd
$7\frac{1}{2}$ yd
$7\frac{1}{2}$ yd

6.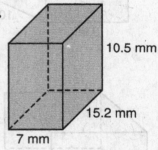
13 m
9 m
5 m

7.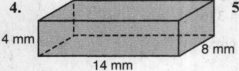
2.4 cm 12.3 cm
16 cm

8.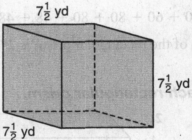
$11\frac{1}{2}$ in.
3 in.
6 in.

9.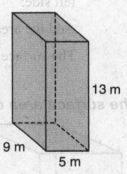
10.5 mm
15.2 mm
7 mm

10. length, 3.2 cm
width, 5 cm
height, 8.6 cm

11. length, 13 yd
width, 12 yd
height, 11 yd

12. length, 15 in.
width, 4 in.
height, 9 in.

13. length, 10.0 m
width, 6.8 m
height, 7.7 m

14. length, 20 mm
width, 16 mm
height, 2 mm

15. length, 24 ft
width, 18 ft
height, 12 ft

Mathematics: Applications and Connections, Course 1

11-1 Study Guide

Integers

An **integer** is any number from the set $\{\ldots, -3, -2, -1, 0, 1, 2, 3, \ldots\}$ where ... means *continues without end*.

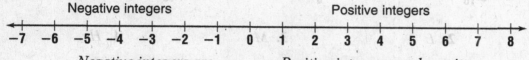

Negative integers   Positive integers

Negative integers are written with a − sign.   *Positive integers can be written with or without a + sign.*

Examples   **1** **Write an integer to show 5 degrees below zero.**

Write: -5

2 **Write an integer to show a 7 degree rise in temperature.**

Write: $+7$ or 7

Opposite integers are the same distance from zero on opposite sides of the number line.

Example   **3** **Write the opposite of $+5$.**

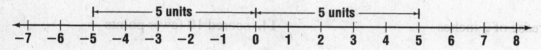

5 units   5 units

The opposite of $+5$ is -5.

Write the integer represented by each letter on the number line.

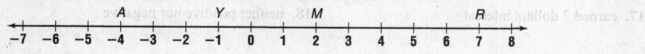

1. M **2.** A **3.** Y **4.** R

Write an integer to describe each situation.

5. 4 feet below sea level   **6.** a gain of 8 points

7. 2 degrees above zero   **8.** a loss of 6 pounds

Write the opposite of each integer.

9. 6   **10.** -2 **11.** 14   **12.** -10

11-1 Practice

Integers

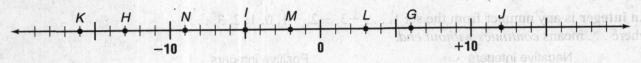

1. *I*   2. *L*   3. *M*   4. *H*

5. *J*   6. *K*   7. *G*   8. *N*

Write an integer to describe each situation.

9. a gain of 5 pounds   10. 4 degrees below normal

11. a loss of 8 yards   12. positive 16

13. an increase of 2 inches   14. scored 10 fewer points

15. negative eighteen   16. 15 feet above sea level

17. earned 7 dollars interest   18. neither positive nor negative

19. bowled 9 pins above average   20. a decrease of 6 members

Write the opposite of each integer.

21. -25 22. 36   23. 54   24. -11

25. 98   26. -47 27. -62 28. 80

29. -73 30. 14   31. 105   32. -29

11-2 Study Guide

Comparing and Ordering Integers

You can use a number line to compare integers. On a number line, the number on the left is always less than the number on the right.

Examples **1** Replace the ◯ in −3 ◯ −7 with <, >, or =.

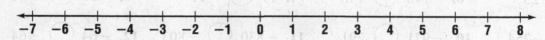

−7 −6 −5 −4 −3 −2 −1 0 1 2 3 4 5 6 7 8

−7 is to the left of −3, so −3 > −7.

2 Order the integers −5, 2, 0, −1 from least to greatest.

Write the integers as they appear on the number line from left to right.

−5, −1, 0, 2

Replace each ◯ with <, >, or = to make a true sentence.

1. −2 ◯ 0   2. 5 ◯ −1   3. −4 ◯ 4   4. −9 ◯ −9

5. −44 ◯ −4   6. −19 ◯ 9   7. −13 ◯ −23   8. 0 ◯ −54

Order each set of integers from least to greatest.

9. −2, 4, 0, −1, 1

10. 0, −4, 4, 7, −6, −5

11. −10, 12, −13, 9, −8, 4

12. 23, 0, 15, −26, −34, −30

13. −7, −19, 19, 0, −25, 30

14. 55, −15, −5, −55, 5, 15

11-2 Practice

Comparing and Ordering Integers

Fill in each ◯ with <, >, or =.

1. -9 ◯ 8 2. 0 ◯ -1 3. -14 ◯ -15 4. $+26$ ◯ 26

5. -32 ◯ 23 6. -148 ◯ 148 7. 19 ◯ -91 8. -67 ◯ -60

9. 245 ◯ -254 10. -971 ◯ 791 11. -830 ◯ -803 12. -64 ◯ -64

13. -57 ◯ -75 14. -33 ◯ 3 15. 169 ◯ -196 16. -200 ◯ -201

Order each set of integers from least to greatest.

17. $-6, 16, -26$ 18. $-213, -231, 132$

19. $5, -3, -11, 9, -7$ 20. $8, -6, 4, -10, -2$

21. $-36, 28, -4, -17, -59$ 22. $-84, 95, -71, -103, -62$

23. $21, -34, 65, -12, 43, 0, -56$ 24. $-22, 2, 0, 202, 22, -222, -2$

25. $-3, -33, 36, -66, 63, -6, 0$ 26. $0, -172, 1, -127, 7, -171, -117$

27. Which is greater, negative 21 or negative 22?

28. Which is greater, 8 degrees above zero or 9 degrees below zero?

29. Which is less, a checkbook balance of -30 dollars or a balance of -25 dollars?

30. Which is greater, negative 98 or positive 89?

Mathematics: Applications and Connections, Course 1

11-3 Study Guide

Adding Integers

You can add integers using models.

Examples

1 **Use counters to find −7 + 2.**

Place 7 negative counters on the mat to represent −7.
Place 2 positive counters on the mat to represent adding 2.
Pair the positive and negative counters.
Remove as many zero pairs as possible.

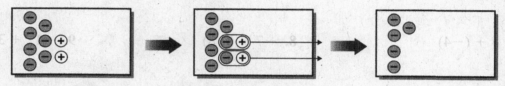

There are 5 negative counters left on the mat.
So, −7 + 2 = −5.

2 **Use counters to find −4 + (−4).**

Place 4 negative counters on the mat to represent −4.
Place 4 more negative counters on the mat to represent adding −4.
Since there are no positive counters, you cannot remove any zero pairs.

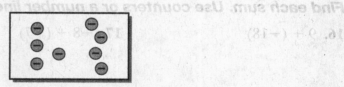

There are 8 negative counters left on the mat.
So, −4 + (−4) = −8.

State whether each sum is positive or negative.

1. −4 + (−2)   2. 5 + (−3)   3. −10 + 7

4. 9 + (−3)   5. 6 + 0   6. −8 + (−1)

Find each sum. Use counters or a number line if necessary.

7. 3 + (−6)   8. −9 + 8   9. −4 + 7

10. 6 + (−6)   11. −8 + (−2)   12. 2 + (−5)

11-3 Practice

Study Guide

11-3

Adding Integers

State whether each sum is positive, negative, or zero.

1. $6 + (-8)$ **2.** $9 + (-3)$ **3.** $-5 + (-4)$

4. $-13 + 7$ **5.** $-2 + 11$ **6.** $10 + (-6)$

7. $4 + (-4)$ **8.** $-7 + (-8)$ **9.** $-12 + 3$

10. $-5 + 14$ **11.** $-10 + (-2)$ **12.** $6 + (-1)$

13. $-8 + 4$ **14.** $7 + (-9)$ **15.** $-3 + 11$

Find each sum. Use counters or a number line if necessary.

16. $9 + (-18)$ **17.** $-8 + (-7)$ **18.** $14 + (-6)$

19. $-12 + 5$ **20.** $-4 + 10$ **21.** $-9 + (-8)$

22. $3 + (-11)$ **23.** $-6 + 13$ **24.** $-12 + 6$

25. $-7 + 16$ **26.** $13 + (-9)$ **27.** $-5 + (-5)$

28. $11 + (-6)$ **29.** $-14 + 9$ **30.** $15 + (-7)$

11-4 Study Guide

Subtracting Integers

You can subtract integers using models.

Examples   **1   Use counters to find $-5 - (-3)$.**

Place 5 negative counters on the mat to represent -5.
Remove 3 negative counters to represent subtracting -3.

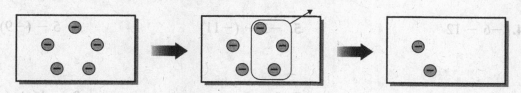

There are 2 negative counters left on the mat.
So, $-5 - (-3) = -2$.

2   Use counters to find $-4 - 3$.

Place 4 negative counters on the mat to represent -4.
To subtract $+3$, you must remove 3 positive counters.
But there are no positive counters on the mat. You must
add 3 zero pairs to the mat. The value of the mat does
not change. Then you can remove 3 positive counters.

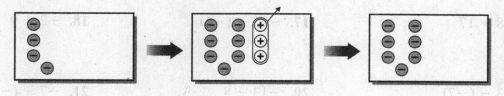

There are 7 negative counters left on the mat.
So, $-4 - 3 = -7$.

Find each difference. Use counters or a number line if necessary.

1. $-8 - (-4)$ **2.** $7 - (-5)$ **3.** $-4 - 2$

4. $-3 - (-5)$ **5.** $6 - (-10)$ **6.** $8 - 5$

7. $-1 - 4$ **8.** $2 - (-2)$ **9.** $-5 - (-1)$

10. $7 - 2$ **11.** $-9 - (-6)$ **12.** $8 - (-2)$

*Mathematics: Applications
and Connections, Course 1*

11-4 Practice

Subtracting Integers

Find each difference. Use counters or a number line if necessary.

1. $4 - 9$

2. $-10 - (-7)$

3. $8 - (-5)$

4. $-6 - 12$

5. $-3 - (-11)$

6. $5 - (-9)$

7. $-8 - 7$

8. $2 - 6$

9. $-16 - (-9)$

10. $4 - (-15)$

11. $-18 - 5$

12. $-6 - 6$

13. $-8 - (-14)$

14. $13 - (-7)$

15. $19 - (-6)$

16. $8 - 17$

17. $-12 - (-4)$

18. $-2 - 9$

19. $3 - (-7)$

20. $-13 - 8$

21. $-7 - (-12)$

22. $3 - 5$

23. $-8 - 6$

24. $-2 - (-2)$

25. $7 - (-4)$

26. $-16 - (-8)$

27. $12 - (-12)$

28. $-3 - 10$

29. $-1 - (-4)$

30. $9 - (-6)$

11-5 Study Guide

Multiplying Integers

Multiplication is repeated addition. You can multiply integers using models.

Examples **1** **Use counters to find $4 \times (-2)$.**

$4 \times (-2)$ means to *put in* 4 sets of 2 negative counters.
Place these counters on the mat.

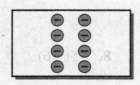

There are 8 negative counters on the mat.
So, $4 \times (-2) = -8$.

2 **Use counters to find $-4(-2)$.**

Since -4 is the opposite of 4, $-4(-2)$ means to
remove 4 sets of 2 negative counters. Add 4 sets of
zero pairs. Then remove 4 sets of 2 negative counters.

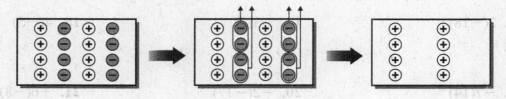

There are 8 positive counters left on the mat.
So, $-4(-2) = 8$.

Find each product. Use counters or a number line if necessary.

1. $3 \times (-3)$ **2.** $-5 \times (-2)$ **3.** $-8 \times (-1)$

4. $-7(8)$ **5.** $9(-3)$ **6.** $-7(-5)$

7. $5(-8)$ **8.** $-10(-4)$ **9.** $-3(4)$

10. $-3(-10)$ **11.** $6(-4)$ **12.** $-7(-7)$

11-5 Practice

Multiplying Integers

Find each product. Use counters or a number line if necessary.

1. $6 \times (-4)$

2. -8×7

3. $-2 \times (-9)$

4. $5(-12)$

5. $-15(-3)$

6. $-4(8)$

7. $9(-7)$

8. $-5(-6)$

9. $3(-16)$

10. $-14(2)$

11. $-4(-4)$

12. $-9(6)$

13. $7(-3)$

14. $-12(-8)$

15. $-5(-15)$

16. $2(-18)$

17. $-3(6)$

18. $4(-5)$

19. $-7(14)$

20. $-2(-17)$

21. $-6(-8)$

22. $4(-13)$

23. $-16(-5)$

24. $-9(12)$

25. $-3(-18)$

26. $7(-15)$

27. $-2(19)$

28. $8(-8)$

29. $-9(11)$

30. $-17(-4)$

Mathematics: Applications and Connections, Course 1

11-6 Study Guide

Dividing Integers

You can use counters to help you divide integers.

Example 1   **Use counters to find −8 ÷ 4.**

Place 8 negative counters on the mat to represent −8. Separate the 8 counters into 4 equal-sized groups.

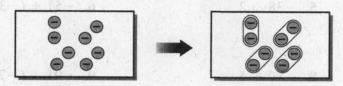

There are 4 groups of 2 negative counters each.
So, −8 ÷ 4 = −2.

You can also use the relationship between multiplication and division to help you divide integers.

Examples 2   **Find −15 ÷ (−3).**

$$5 \times (-3) = -15, \text{ so } -15 \div (-3) = 5.$$

3   **Find 18 ÷ (−2).**

$$-9 \times (-2) = 18, \text{ so } 18 \div (-2) = -9.$$

When you divide a negative integer and a positive integer, the quotient is negative. When you divide two negative integers, the quotient is positive.

Find each quotient. Use counters or patterns if necessary.

1. 27 ÷ (−3)

2. −40 ÷ (−8)

3. −36 ÷ 6

4. −72 ÷ 8

5. 56 ÷ (−7)

6. −81 ÷ (−9)

7. 144 ÷ (−12)

8. −100 ÷ (−5)

9. −84 ÷ 4

10. −93 ÷ (−3)

11. 77 ÷ (−7)

12. −64 ÷ (−8)

11-6 Practice

Dividing Integers

Find each quotient. Use counters or patterns if necessary.

1. $56 \div (-7)$

2. $-45 \div 9$

3. $-72 \div (-6)$

4. $60 \div (-4)$

5. $-38 \div 2$

6. $-51 \div (-3)$

7. $48 \div (-8)$

8. $-80 \div 5$

9. $-91 \div (-7)$

10. $36 \div (-9)$

11. $-42 \div 3$

12. $-54 \div (-6)$

13. $110 \div (-10)$

14. $-126 \div 7$

15. $-35 \div (-5)$

16. $-27 \div 9$

17. $104 \div (-8)$

18. $-32 \div (-16)$

19. $-68 \div 4$

20. $-120 \div (-8)$

21. $-84 \div (-6)$

22. $132 \div (-11)$

23. $49 \div (-7)$

24. $-18 \div 2$

25. $30 \div (-3)$

26. $-75 \div (-15)$

27. $-28 \div 14$

28. $-116 \div (-29)$

29. $256 \div (-32)$

30. $-144 \div 24$

Mathematics: Applications and Connections, Course 1

11-7 Study Guide

Integration: Geometry
The Coordinate System

A horizontal number line and a vertical number line meet at their zero points to form a **coordinate system.** The horizontal line is called the **x-axis.** The vertical line is called the **y-axis.** The location of a point in the coordinate system can be named using an ordered pair of numbers.

(x, y)

x-coordinate ——————⌐‾⌐—————— y-coordinate

Examples

1 **Name the ordered pair for point P.**

Start at 0. Move right along the x-axis until you are directly above point P. Then move parallel to the y-axis until you reach point P. Since you moved 4 units to the right and 3 units down, the ordered pair for point P is (4, −3).

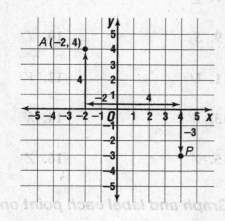

2 **Graph A(−2, 4).**

Start at 0. Move 2 units left on the x-axis. Then move 4 units up parallel to the y-axis to locate the point.

Name the ordered pair for each point.

1. B   2. L

3. F   4. H

5. K   6. E

7. J   8. C

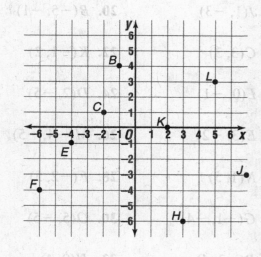

Graph and label each point.

9. $A(-5, 5)$ 10. $M(2, 4)$

11. $G(0, -5)$ 12. $D(-6, 0)$

13. $N(6, 6)$ 14. $I(4, -3)$

Mathematics: Applications and Connections, Course 1

11-7 Practice

Integration: Geometry
The Coordinate System

Name the ordered pair for each point.

1. K

2. L

3. M

4. N

5. O

6. P

7. Q

8. R

9. S

10. T

11. U

12. V

13. W

14. X

15. Y

16. Z

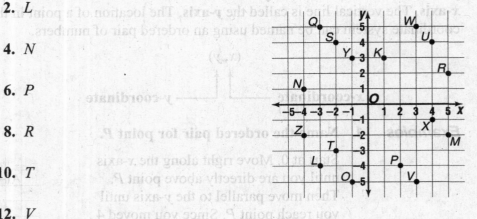

Graph and label each point on the coordinate grid below.

17. $A(-5, 2)$

18. $I(2, 1)$

19. $J(1, -3)$

20. $B(-5, -1)$

21. $C(3, 3)$

22. $K(-1, 2)$

23. $L(0, -1)$

24. $D(2, -5)$

25. $E(3, -2)$

26. $M(-4, -5)$

27. $N(1, 5)$

28. $F(-2, 5)$

29. $G(-1, -4)$

30. $O(5, -5)$

31. $P(-3, 4)$

32. $H(0, 4)$

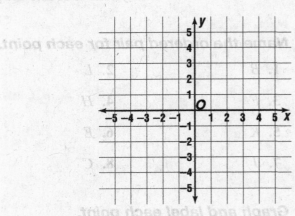

Mathematics: Applications and Connections, Course 1

11-8 Study Guide

Integration: Geometry
Graphing Transformations

A **transformation** is the movement of a figure. There are several kinds of transformations. A **translation** is a slide. A **reflection** is a flip.

Examples

1 The vertices of triangle *LMN* are *L*(1, 1), *M*(3, 4), and *N*(5, 2). On a coordinate grid, draw triangle *LMN* and its translation image that is 1 unit to the left and 3 units down.

• Graph and label the vertices of triangle *LMN*.
• Draw triangle *LMN*.
• Translate each vertex 1 unit to the left and 3 units down. The coordinates of the new vertices are *L'*(0, −2), *M'*(2, 1), and *N'*(4, −1).
• Label the new vertices *L'*, *M'*, and *N'*.
• Then draw triangle *L'M'N'*.

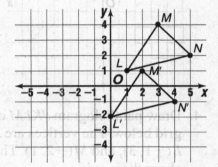

2 The vertices of triangle *QRS* are *Q*(−2, 4), *R*(0, 1), and *S*(−3, −1). On a coordinate grid, draw triangle *QRS* and its reflection image over the *y*-axis.

• Graph and label the vertices of triangle *QRS*.
• Draw triangle *QRS*.
• Reflect the triangle by flipping it onto the other side of the *y*-axis. To be a mirror image, each new vertex must be the same distance from the *y*-axis as its corresponding vertex. The coordinates of the new vertices are *Q'*(2, 4), *R'*(0, 1), and *R'*(3, −1).
• Label the new vertices *Q'*, *R'*, and *S'*.
• Then draw triangle *Q'R'S'*.

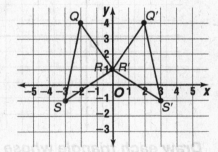

The vertices of parallelogram ABCD are A(−1, 1), B(1, 4), C(4, 4), D(2, 1). On the coordinate grid, draw parallelogram ABCD and each transformation image.

1. translation image that is 5 units to the left and 1 unit down

2. reflection image over the *x*-axis

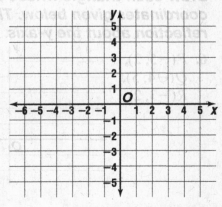

83

Mathematics: Applications and Connections, Course 1

11-8 Practice

11-8 Study Guide

Integration: Geometry
Graphing Transformations

Tell whether each transformation is a translation or a reflection.

1.

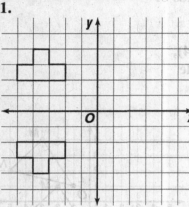

2.

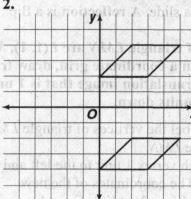

3.

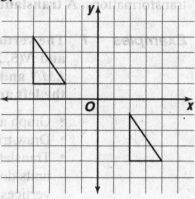

4. Draw parallelogram *JKLM* on the coordinate grid below. The vertices are $J(-4, 1)$, $K(-3, 3)$, $L(-1, 3)$, and $M(-2, 1)$. Then draw its reflection about the *x*-axis.

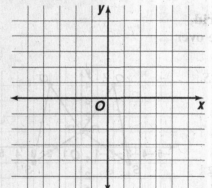

5. Name the coordinates of the vertices for the figure below. On the coordinate grid, draw a translation image 6 units to the right and 5 units down. What are the new coordinates of the vertices of the translation image?

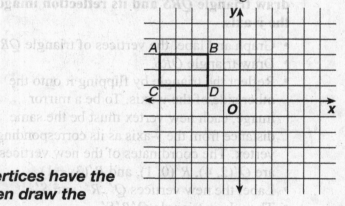

Draw each triangle whose vertices have the coordinates given below. Then draw the reflection about the *y*-axis.

6. $P(-5, 2)$,
   $Q(-4, 5)$,
   $R(-3, 2)$

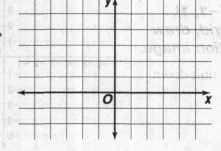

7. $S(1, -2)$,
   $T(4, -2)$,
   $R(4, -5)$

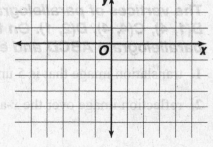

12-1 Study Guide

Solving Addition Equations

To **solve an equation** means to find a value for the variable that makes the equation true.

Example   **Use cups and counters to solve $b + (-3) = 5$.**

Use a cup to represent b. Add 3 negative counters on the left side of the mat to represent -3. Place 5 positive counters on the right side of the mat to represent $+5$.

$b + (-3) = +5$

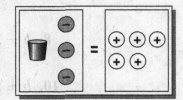

To get the cup by itself, you need to remove 3 negative counters from each side. Since there are no negative counters on the right side of the mat, add 3 positive counters to each side to make 3 zero pairs on the left side of the mat. Then remove the zero pairs.

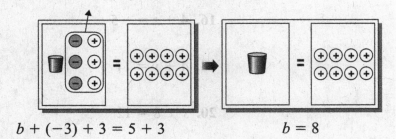

$b + (-3) + 3 = 5 + 3$ $b = 8$

Solve each equation. Use cups and counters if necessary.

1. $k + 5 = 6$ 2. $a + 4 = -5$ 3. $m + (-8) = -15$

4. $v + (-1) = -7$ 5. $x + 3 = -3$ 6. $c + 5 = 10$

7. $4 + d = 7$ 8. $g + (-1) = -3$ 9. $6 + t = -2$

10. If $z + 7 = -12$, what is the value of z?   11. Find the value of v if $15 + v = -2$.

12-1 Practice

12-1 Study Guide

Solving Addition Equations

Solve each equation. Use cups and counters if necessary.

1. $a + 6 = 11$

2. $s + (-3) = -10$

3. $m + 8 = -2$

4. $5 + p = -7$

5. $c + (-10) = -6$

6. $t + 9 = -5$

7. $7 + y = 16$

8. $h + 2 = -6$

9. $2 + k = -9$

10. $b + 5 = -12$

11. $n + 2 = 4$

12. $9 + r = -7$

13. $j + 2 = -1$

14. $y + 6 = 14$

15. $e + 7 = -6$

16. $4 + z = -5$

17. $x + 3 = -3$

18. $4 + v = 11$

19. $q + 6 = -9$

20. $s + 8 = 12$

21. $8 + p = -3$

22. $x + 6 = -8$

23. $4 + m = -12$

24. $w + 8 = -7$

25. $f + (-4) = 9$

26. $d + 9 = -3$

27. $g + (-8) = 9$

28. $j + 2 = 7$

29. $u + (-5) = 5$

30. $r + (-5) = -2$

12-2 Study Guide

Solving Subtraction Equations

To **solve an equation** means to find a value for the variable that makes the equation true.

Example   **Use cups and counters to solve $m - 2 = 5$.**

Rewrite as an addition equation.
$$m - 2 = 5 \longrightarrow m + (-2) = 5$$

Use a cup to represent m. Add 2 negative counters on the left side of the mat to represent -2. Place 5 positive counters on the right side of the mat to represent $+5$.

$m + (-2) = 5$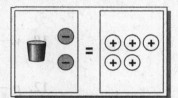

To get the cup by itself, you need to remove 2 negative counters from each side. Since there are no negative counters on the right side of the mat, add 2 positive counters to each side to make 2 zero pairs on the left side of the mat. Then remove the zero pairs.

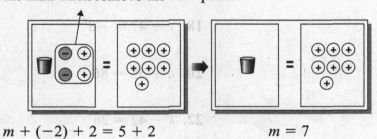

$m + (-2) + 2 = 5 + 2$ $m = 7$

Solve each equation. Use cups and counters if necessary.

1. $a - 3 = 5$ 2. $r - 5 = -8$ 3. $h - 2 = 0$

4. $j - 8 = 8$ 5. $77 = d - 12$ 6. $n - 20 = -5$

7. $r - (-10) = 15$ 8. $y - (-1) = 4$ 9. $n - (-23) = -55$

10. If $h - 9 = 25$, what is the value of h?   11. Find the value of c if $c - 19 = -10$.

Mathematics: Applications and Connections, Course 1

12-2 Practice

Solving Subtraction Equations

Solve each equation. Use cups and counters if necessary.

1. $h - (-2) = 6$

2. $v - 7 = -4$

3. $a - (-6) = -5$

4. $r - (-3) = -8$

5. $j - (-8) = 5$

6. $x - 8 = -9$

7. $c - 26 = 45$

8. $z - (-57) = -39$

9. $n - 38 = -19$

10. $w - 23 = 77$

11. $f - (-26) = 41$

12. $p - 47 = 22$

13. $g - 82 = -63$

14. $t - 14 = 87$

15. $q - 53 = 27$

16. $b - 48 = 14$

17. $k - 7 = -2$

18. $y - 47 = -8$

19. $t - 33 = -51$

20. $a - 35 = 86$

21. $n - 84 = 16$

22. $k - 42 = 26$

23. $x - 33 = -52$

24. $y - 63 = -19$

25. $d - 47 = 42$

26. $r - 47 = 84$

27. $b - 42 = 63$

28. $y - 18 = -47$

29. $j - 92 = -20$

30. $s - 26 = -99$

12-3 Study Guide

Solving Multiplication and Division Equations

When the variable of an equation is multiplied by a number, divide each side of the equation by the number to get the variable by itself.

Example 1 Use cups and counters to solve $3n = -9$.

Use three cups to represent $3n$. Place 9 negative counters on the right side of the mat to represent -9.

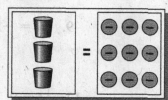

$$3n = -9$$

Undo the multiplication by dividing each side by 3. Show division by 3 by forming 2 equal groups on each side of the mat.

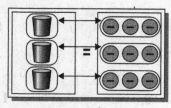

$$\frac{3n}{3} = \frac{-9}{3} \longrightarrow n = -3$$

When the variable of an equation is divided by a number, multiply each side of the equation by that number to get the variable by itself.

Example 2 Use cups and counters to solve $\frac{1}{3}p = 2$.

Use a cup that is about $\frac{1}{3}$ full to represent $\frac{1}{3}n$. Place 2 positive counters on the right side of the mat to represent $+2$.

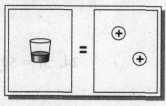

$$\frac{1}{3}p = 2$$

Undo the division by multiplying each side by 3. Place 3 sets of 2 positive counters on the right side of the mat.

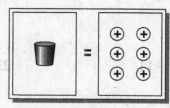

$$3\left(\frac{1}{3}p\right) = 3(2) \longrightarrow p = 6$$

Solve each equation. Use cups and counters if necessary.

1. $5t = 15$

2. $2k = -14$

3. $4p = -16$

4. $3m = -15$

5. $6n = -24$

6. $\frac{1}{2}c = 2$

7. $\frac{1}{3}r = 4$

8. $\frac{1}{2}f = -3$

9. $\frac{1}{4}y = -1$

Mathematics: Applications and Connections, Course 1

12-3 Practice

Solving Multiplication and Division Equations

Solve each equation. Use cups and counters if necessary.

1. $9k = 54$

2. $7r = -35$

3. $\frac{1}{2}y = 6$

4. $\frac{1}{3}g = -12$

5. $4a = -28$

6. $\frac{1}{8}m = -6$

7. $\frac{1}{5}w = 2$

8. $6s = 42$

9. $\frac{1}{4}h = -5$

10. $\frac{1}{9}x = -8$

11. $3p = 27$

12. $\frac{1}{7}t = 9$

13. $5d = -30$

14. $\frac{1}{6}j = -12$

15. $8n = -64$

16. $2c = 28$

17. $\frac{1}{5}k = -9$

18. $7f = -91$

19. $\frac{1}{3}z = 45$

20. $4q = -48$

21. $\frac{1}{9}b = 2$

22. $\frac{1}{8}e = -11$

23. $6u = 3$

24. $5i = 50$

25. $\frac{1}{7}y = -7$

26. $3a = -48$

27. $\frac{1}{2}r = 20$

28. $\frac{1}{4}s = -8$

29. $9p = -108$

30. $\frac{1}{6}x = 6$

12-4 Study Guide

Solving Two-Step Equations

To solve a two-step equation, undo the addition or subtraction. Then undo the multiplication or division.

Example   **Use cups and counters to solve $3b - 2 = -5$.**

Rewrite as an addition equation.

$$3b - 2 = -5 \longrightarrow 3b + (-2) = -5$$

Place 3 cups and 2 negative counters on the left side of the mat to represent $3b - 2$. Place 5 negative counters on the right side of the mat to represent -5.

$3b - 2 = -5$

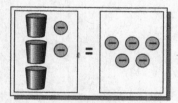

To get the cups by themselves, you need to remove 2 negative counters from each side.

$3b + (-2) - (-2) = -5 - (-2)$

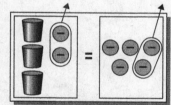

Now the equation is $3b = -3$. Undo the multiplication by dividing each side by 3. Show division by forming 3 equal groups on each side of the mat.

$\frac{3b}{3} = \frac{-3}{3} \longrightarrow b = -1$

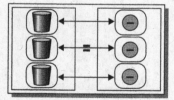

Solve each equation.

1. $5x + 3 = 23$

2. $3a - 14 = 4$

3. $3y + 5 = -19$

4. $5c + 6 = -29$

5. $42 = 18 - 4v$

6. $8 - 5w = -37$

7. Five less than 4 times a number is nineteen. What is the number?

       Mathematics: Applications and Connections, Course 1

12-4 Practice

Solving Two-Step Equations

Solve each equation.

1. $2x - 5 = 3$

2. $4t + 5 = 37$

3. $7a + 6 = -36$

4. $-8g + 7 = 47$

5. $-3c - 9 = -24$

6. $5k - 7 = -52$

7. $5s + 4 = -71$

8. $6y - 30 = 18$

9. $7p - 70 = -21$

10. $-4m + 5 = 25$

11. $6z - 1 = 23$

12. $2b + 12 = -6$

13. $\frac{c}{5} + 6 = -2$

14. $\frac{d}{4} - 8 = -5$

15. $\frac{a}{7} + 9 = 3$

16. $2v - 3 = 9$

17. $-3x - 4 = 2$

18. $6y + 10 = -8$

19. $2w - 7 = -3$

20. $-4t + 15 = -1$

21. $-9k - 3 = -21$

22. $-6h + 7 = -17$

23. $5n + 3 = -17$

24. $12g + 48 = 72$

25. $7v + 3 = -46$

26. $\frac{b}{5} - 19 = 11$

27. $\frac{d}{-4} + 9 = 2$

28. $\frac{q}{8} - 1 = 12$

29. $3 - \frac{a}{6} = 7$

30. $5 + \frac{n}{-3} = 12$

12-5 Study Guide

Functions

A function connects a number *n*, the input, to another number, the output, by a rule.

Example   Replace *n* with −4, −2, 0, 4 in the rule 2*n* − 1.

input (*n*)	output (2*n* − 1)
−4	2(−4) − 1 = −9
−2	2(−2) − 1 = −5
0	2(0) − 1 = −1
4	2(4) − 1 = 7

Complete each function table.

1.

input (*n*)	output (*n* + 4)
−2	
0	
1	
3	

2.

input (*n*)	output (3*n*)
−1	
2	
3	
5	

3.

input (*n*)	output [*n* + (−1)]
−5	
−1	
4	
8	

4.

input (*n*)	output $\left(\frac{n}{2} + 2\right)$
−6	
0	
2	
6	

Find the rule for each function table.

5.

n	
−4	−2
0	2
2	4
5	7

6.

n	
−3	−6
−1	−4
1	−2
3	0

7.

n	
−2	−1
0	0
2	1
4	2

12-5 Practice

Functions

Complete each function table.

1.

input (n)	output (7n)
6	
8	
9	

2.

input (n)	output (n + 9)
−5	
0	
7	

3.

input (n)	output (n − 8)
−4	
0	
10	

4.

input (n)	output $\left(\frac{n}{3} - 2\right)$
−6	
0	
9	

5. If a function rule is $8x + 5$, what is the output for $x = -3$?

6. If a function rule is $-4\left(\frac{x}{2}\right)$, what is the output for $x = 1.2$?

Find the rule for each function table.

7.

input (n)	output
−1	3
0	4
2	6

8.

input (n)	output
−3	−15
0	−12
4	−8

9.

input (n)	output
−12	−2
0	0
6	1

10.

input (n)	output
0	2
1	5
2	8

Study Guide

Graphing Functions

To graph a function, let the input number be the *x*-coordinate. Let the output number be the *y*-coordinate.

Example **Make a function table for the rule 2x. Then graph the function.**

Choose input values. Find output values. Then graph the ordered pairs and draw a line.

input (x)	output (2x)	ordered pairs
−2	−4	(−2, −4)
0	0	(0, 0)
2	4	(2, 4)
3	6	(3, 6)

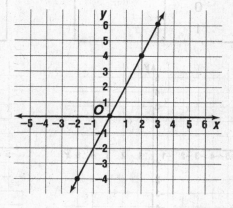

Complete each function table. Then graph the function.

1.

input (n)	output $\left(\frac{n}{3}\right)$	ordered pairs
−3		
0		
6		

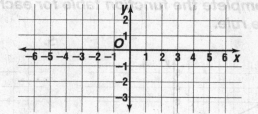

2.

input (n)	output (n + 1)	ordered pairs
−3		
−1		
2		

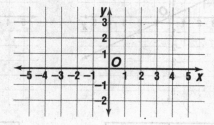

3.

input (n)	output (n − 2)	ordered pairs
0		
2		
5		

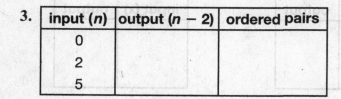
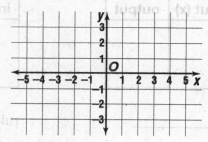

Mathematics: Applications and Connections, Course 1

12-6 Practice

Graphing Functions

Complete each function table. Then graph each function.

1.

input (x)	output (2x)
−2	
0	
1	

2.

input (x)	output (x + 4)
−5	
−3	
0	

3.

input (x)	output (3 − x)
2	
3	
4	

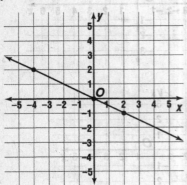

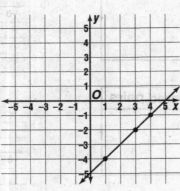

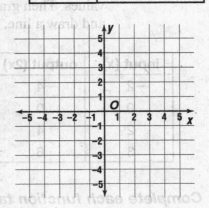

Complete the function table for each graph. Then determine the rule.

4.

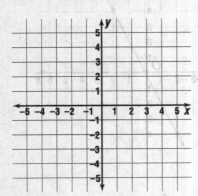

input (x)	output

rule: _____

5.

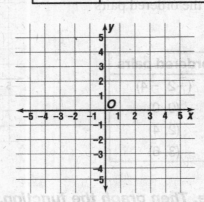

input (x)	output

rule: _____

6.

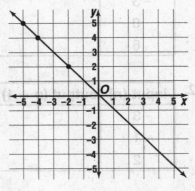

input (x)	output

rule: _____

Mathematics: Applications and Connections, Course 1

13-1 Study Guide

Probability

The **probability** of an event is the ratio of the number of ways the event can occur to the number of possible outcomes.

$$P(\text{event}) = \frac{\text{number of ways the event can occur}}{\text{number of possible outcomes}}$$

Examples

1 **On the spinner below, there are eight equally likely outcomes. Find the probability of spinning a number less than 3.**

Numbers less than 3 are 1 and 2.
There are 8 possible outcomes.

$P(\text{less than 3}) = \frac{2}{8}$ or $\frac{1}{4}$

2 **Find $P(\text{greater than 10})$.**

$P(\text{greater than 10}) = \frac{0}{8}$ or 0

3 **Find $P(\text{less than 9})$.**

$P(\text{less than 9}) = \frac{8}{8}$ or 1

Suppose you choose one of the cards shown without looking. Find the probability of each event.

1. $P(12)$ **2.** $P(\text{even})$ **3.** $P(\text{2 digits})$

4. $P(\text{prime})$ **5.** $P(\text{odd})$ **6.** $P(\text{less than 8})$

7. $P(\text{greater than 40})$ **8.** $P(\text{divisible by 3})$

John has 15 baseball caps. 4 are red, 6 are blue, 3 are yellow, and 2 are white. If he chooses one without looking, find each probability.

9. $P(\text{yellow})$ **10.** $P(\text{red or blue})$ **11.** $P(\text{black})$

12. $P(\text{white})$ **13.** $P(\text{red or white})$ **14.** $P(\text{yellow or white})$

13-1 Practice

Probability

Jared keeps his socks in random order in his top dresser drawer. There are two brown socks, eight black socks, four gray socks, and two blue socks in his drawer. Jared reaches into the drawer and, without looking, grabs one sock. Find the probability of each event.

1. $P(\text{gray})$

2. $P(\text{blue})$

3. $P(\text{black})$

4. $P(\text{white})$

5. $P(\text{brown or black})$

6. $P(\text{gray or blue})$

A set of 52 playing cards contains 4 different suits of 13 cards each. Hearts and diamonds are red; spades and clubs are black. Each suit contains cards numbered 2 through 10, a jack, queen, king, and ace. It is equally likely to choose any one card. Find the probability of each event.

7. $P(\text{red})$

8. $P(\text{clubs})$

9. $P(\text{ace})$

10. $P(\text{jack of diamonds})$

11. $P(\text{black 10})$

12. $P(\text{red king or queen})$

13. $P(\text{black 2, 3, or 4})$

14. $P(\text{red 1})$

Mrs. Phipps found 10 identical cans without labels in her cupboard. She knew that she originally had two cans of peas, five cans of corn, one can of carrots, and two cans of beets. She opens one can. Find the probability of each event.

15. $P(\text{carrots})$

16. $P(\text{corn})$

17. $P(\text{beans})$

18. $P(\text{peas})$

19. $P(\text{corn or beets})$

20. $P(\text{carrots or peas})$

13-2

Study Guide

Integration: Statistics
Making Predictions Using Samples

Data gathered by surveying a random sample of the population may be used to make predictions about the entire population.

Example   Joyce surveyed every tenth person entering the school to determine whether they would prefer attending a rock concert or a dance. Of the 60 students surveyed, 35 said they preferred a rock concert, and 25 said they preferred a dance. If 800 students attend the school, predict how many would prefer a rock concert.

$\frac{35}{60}$ or 58% of those surveyed said they preferred a concert.

58% of 800 = 0.58 × 800
            = 464

About 464 students would prefer a rock concert.

Solve.

1. In a random sample of 600 batteries, 3 were found to be defective. If a factory produces 7,000 batteries each day, predict the number that are defective.

2. Wilson has made 8 out of the last 20 free throws he has attempted. What is the probability that he will make the next free throw?

3. The graph shows who has bought tickets so far. What is the probability that a senior citizen will buy the next ticket?

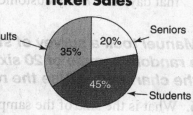

Ticket Sales

Adults 35% — Seniors 20% — Students 45%

4. In a poll of 200 people, 82 said they would vote for Peterson for mayor, 106 said they would vote for Sanderson, and 12 were undecided. If 5,200 people vote in the election, predict the number that will vote for Peterson.

5. In a survey, 35% of the students said they would buy a hot dog at a football game, 45% said they would buy a hamburger, and 85% said they would buy a soft drink. If 350 students are expected to attend a football game, how many hot dogs, hamburgers, and soft drinks should be ordered for the concession stand?

       91   *Mathematics: Applications and Connections,* Course 1

13-2 Practice

Integration: Statistics
Making Predictions Using Samples

Tell whether each of the following is a random sample. Explain your answer.

<u>Type of Survey</u>	<u>Survey Location</u>

1. favorite brand of sneaker   a specific sneaker outlet store

2. students' favorite pizza topping   one student from each homeroom

3. favorite vacation spot   people at the beach in the summer

4. preferred means of traveling to work   people in a weekday traffic jam

5. In a random survey of 50 customers in a supermarket, Bonnie found 16 that drive mini vans. If there were 225 customers in the supermarket that day, how many customers are likely to drive a mini van?

Manuel took a survey of students' favorite core subjects from a random sample of 20 sixth graders. The results are shown in the chart below. Use the results to answer each question.

6. What is the size of the sample?

7. What is the probability that a 6th grader prefers

     a. math?   **b.** science?

8. If there are 100 sixth grade students, what percent prefers

     a. social studies?   **b.** language arts?

6th Graders' Favorite Subjects	
Math	8
Language Arts	3
Science	5
Social Studies	4

Mathematics: Applications and Connections, Course 1

13-3 Study Guide

Integration: Geometry
Probability and Area

You can use *area ratios* to find probabilities.

Example   Suppose you threw 100 darts at this dart board. If all darts hit the board and are equally likely to land on any part of the board, how many would you expect to land in the shaded region?

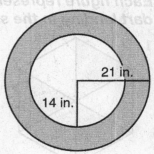

The area of the shaded region is the area of the large circle minus the area of the small circle.

A(large circle) $= \pi r^2$ A(small circle) $= \pi r^2$

$\qquad\qquad = \frac{22}{7} \times 21^2$ $\qquad\qquad = \frac{22}{7} \times 14^2$

$\qquad\qquad = \frac{22}{7} \times 441$ $\qquad\qquad = \frac{22}{7} \times 196$

$\qquad\qquad = 1,386$ $\qquad\qquad = 616$

Shaded area $= 1,386 - 616$ or 770 square inches

Use a proportion to find how many out of 100 darts will probably land in the shaded region.

area of shaded region $\longrightarrow$ $\frac{770}{1,386} = \frac{x}{100}$ $\longleftarrow$ number of darts in shaded region
area of whole region $\longrightarrow$ $\longleftarrow$ total number of darts

$\qquad\qquad\qquad 1,386x = 77,000$
$\qquad\qquad\qquad\qquad x = 55.6$

About 56 out of 100 darts can be expected to land in the shaded region.

Suppose you threw 100 darts at each dart board below. How many would you expect to land in each shaded region?

1.
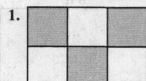

2.
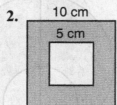

3.
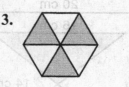

4.
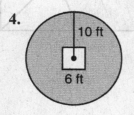

5.
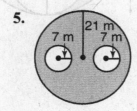

6.
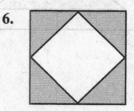

Mathematics: Applications and Connections, Course 1

13-3 Practice

Integration: Geometry
Probability and Area

Each figure represents a dartboard. Find the probability of a dart landing in the shaded area.

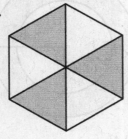 1.

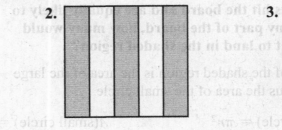 2.

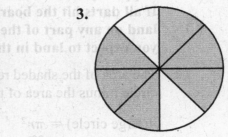 3.

Suppose you throw 100 darts at each dartboard below. How many darts would you expect to land in each shaded area?

 4.

 5.

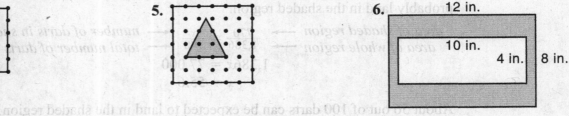 6. 12 in. 10 in. 4 in. 8 in.

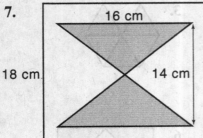 7. 20 cm 16 cm 18 cm 14 cm

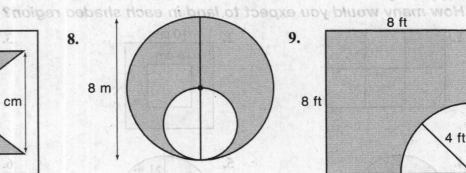 8. 8 m

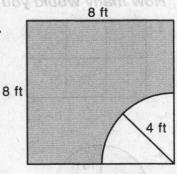 9. 8 ft 8 ft 4 ft

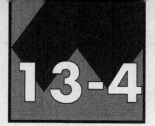

13-4 **Study Guide**

Finding Outcomes

One way to show all of the possible outcomes is to organize data in a *tree diagram*.

Example   Ernie can order a small, medium, or large pizza with thick or thin crust. How many possible ways can he order?

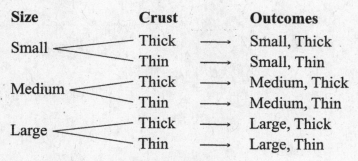

Size	Crust	Outcomes
Small	Thick →	Small, Thick
	Thin →	Small, Thin
Medium	Thick →	Medium, Thick
	Thin →	Medium, Thin
Large	Thick →	Large, Thick
	Thin →	Large, Thin

There are 6 ways Ernie can order.

For each situation, draw a tree diagram to show the sample space.

1. Spin each spinner once.

2. José, Kara, and Beth are running for class president. Tony, Lou, and Fay are running for vice-president.

3. You can buy toothpaste in small, medium, or large size and regular or mint flavor.

4. Toss a dime and a quarter.

Mathematics: Applications and Connections, Course 1

13-4 Practice

13-4 Practice

Finding Outcomes

For each situation, draw a tree diagram to show all of the possible outcomes.

1. Each spinner is spun once.

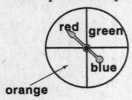

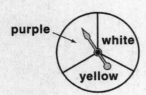

2. The sweaters on sale come in 3 styles: pullover, cardigan, and turtleneck. They come in 3 colors: white, black, or tan.

3. You can have a choice of chocolate, vanilla, or strawberry frozen yogurt in a waffle or a sugar cone.

13-5 Study Guide

Name_____ Date_____

Probability of Independent Events

If the outcome of one event does not affect the outcome of a second event, the two events are **independent**.

The probability of two independent events, A and B, is equal to the probability of event A times the probability of event B.

$$P(A \text{ and } B) = P(A) \times P(B)$$

Example **Suppose you spin each of these two spinners. What is the probability of spinning an even number and a vowel?**

$P(\text{even}) = \frac{1}{2}$ ← *number of ways to spin even*
$\phantom{P(\text{even}) = \frac{1}{2}}$ ← *number of possible outcomes*

$P(\text{vowel}) = \frac{1}{5}$ ← *number of ways to spin a vowel*
$\phantom{P(\text{vowel}) = \frac{1}{5}}$ ← *number of possible outcomes*

$P(\text{even, vowel}) = \frac{1}{2} \times \frac{1}{5} \text{ or } \frac{1}{10}$

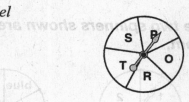

The two spinners shown above are spun. Find the probability of each event.

1. $P(6, P)$

2. $P(\text{less than 4, consonant})$

3. $P(\text{odd, S})$

4. $P(5, \text{consonant})$

5. $P(\text{greater than 8, T})$

6. $P(\text{less than 7, vowel})$

A quarter and a dime are tossed. Find the probability of each event.

7. $P(T, H)$

8. $P(\text{both the same})$

9. $P(T, T)$

Suppose you write each letter of your first and last names on a separate index card and select one letter from each name without looking. Find the probability of each event.

10. $P(\text{vowel, vowel})$

11. $P(\text{consonant, vowel})$

12. $P(M, E)$

Mathematics: Applications and Connections, Course 1

13-5 Practice

Probability of Independent Events

Solve.

1. A bakery shop offers breakfast specials consisting of a muffin and a beverage. There are five kinds of muffins to choose from: corn, blueberry, honey-bran, cranberry-orange, and banana-nut. The beverages available are coffee, tea, orange juice, or milk. Each choice is equally likely.

   a. What is the probability of choosing a blueberry muffin?

   b. What is the probability of choosing orange juice?

   c. Find P(blueberry muffin and orange juice).

The two spinners shown are spun. Find the probability of each event.

 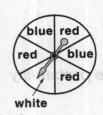

white

2. P(1 and white)

3. P(3 and red)

4. P(2 and blue)

5. P(odd and red)

6. P(odd and blue)

7. P(4 and white)

8. P(even and any color other than white)

Suppose you toss a coin and pick a card from a pile of 18 cards, each printed with a letter from the phrase "MATHEMATICS IS FOR ME." Find the probability of each of the following.

9. P(heads and M)

10. P(tails and F)

11. P(tails and T)

12. P(heads and vowel)

13. P(tails and consonant)

14. P(heads and N)